The Biochemistry and Physiology of *Tetrahymena*

CELL BIOLOGY: A Series of Monographs

EDITORS

D. E. BUETOW

Department of Physiology
and Biophysics
University of Illinois
Urbana, Illinois

I. L. CAMERON

Department of Anatomy
University of Texas
Medical School at San Antonio
San Antonio, Texas

G. M. PADILLA

Department of Physiology and Pharmacology
Duke University Medical Center
Durham, North Carolina

G. M. Padilla, G. L. Whitson, and I. L. Cameron (editors). THE CELL CYCLE: *Gene–Enzyme Interactions*, 1969

A. M. Zimmerman (editor). HIGH PRESSURE EFFECTS ON CELLULAR PROCESSES, 1970

I. L. Cameron and J. D. Thrasher (editors). CELLULAR AND MOLECULAR RENEWAL IN THE MAMMALIAN BODY, 1971

I. L. Cameron, G. M. Padilla, and A. M. Zimmerman (editors). DEVELOPMENTAL ASPECTS OF THE CELL CYCLE, 1971

P. F. Smith. THE BIOLOGY OF MYCOPLASMAS, 1971

Gary L. Whitson. CONCEPTS IN RADIATION CELL BIOLOGY, 1972

Donald L. Hill. THE BIOCHEMISTRY AND PHYSIOLOGY OF *TETRAHYMENA*, 1972

In preparation

Kwang W. Jeon. THE BIOLOGY OF *AMOEBA*

The Biochemistry and Physiology of *Tetrahymena*

Donald L. Hill

KETTERING-MEYER LABORATORY
SOUTHERN RESEARCH INSTITUTE
BIRMINGHAM, ALABAMA

ACADEMIC PRESS New York and London 1972

ACADEMIC PRESS, INC.
111 Fifth Avenue, New York, New York 10003

United Kingdom Edition published by
ACADEMIC PRESS, INC. (LONDON) LTD.
24/28 Oval Road, London NW1 7DD

Library of Congress Catalog Card Number: 75-182618

PRINTED IN THE UNITED STATES OF AMERICA

Contents

Foreword

"For the future development of protozoan biochemistry, it seems of utmost importance that an atmosphere develop in which more biochemists may feel, without external pressure, that many problems of the biochemistry of Protozoa are now ripe for further investigations and that the Protozoa are quite ready to respond to their love and interest." These words of André Lwoff appeared just two decades ago in his Introduction to Volume I of "Biochemistry and Physiology of Protozoa," which he edited. The viewpoint expressed is enormously important. Unfortunately there still are very few protozoan types "domesticated" for biochemical use. A most famous and useful genus, *Tetrahymena*, has contributed most of our biochemical information about the phylum Protozoa. *Tetrahymena* (under another name) was first grown in pure (axenic) culture by Dr. Lwoff in 1922, and thus Dr. Hill may be said to have presented us, in this volume, with a Golden Anniversary Tribute to the organism.

After having spent many years of my life pondering the molecular events (some unique and some mundane) of this marvelous microorganism, it was personally satisfying to be asked by Dr. Hill to read the original manuscript and to write a Foreword to the book. Bringing together the material which appears in this volume was a herculean task and truly a work of love by a scientist who has himself worked effectively with *Tetrahymena* and who has contributed many communications regarding its enzymic equipment.

I congratulate Dr. Hill on a masterly compilation and review of the literature on the life activities of a star performer (from the biochemical standpoint, *the* star performer) of a huge phylum. All biochemists and biologists, in general, will find value in this volume. I commend it to my scientific colleagues.

G. W. Kidder
Amherst, Massachusetts

Preface

This book presents a complete review of the literature covering the physiology and biochemistry of the ciliate genus *Tetrahymena,* of which *T. pyriformis* is the most studied species. I have considered *Tetrahymena* as a whole animal and have tried to integrate all of the pertinent information into a form understandable to those with a basic knowledge of biochemistry or physiology. The comparative aspects of the subject have not been stressed, although I have noted interesting similarities and differences between this organism and others.

For those who are not too familiar with *Tetrahymena,* Chapter 1 provides basic information about this organism that is found in almost any body of water and is so unusual that one can debate quite rationally as to whether it is an animal or a plant. The other chapters are restricted to specific subjects, although there is some overlap. Fortunately, nearly all of the facts about *Tetrahymena* fit readily into one of the ten chapters. Placement of the references at the end of each chapter can serve an investigator interested in only one aspect of the organism since each chapter can be considered an independent entity presenting a review of a particular area complete with references.

The contribution of *Tetrahymena* to scientific knowledge is great. The ciliate has been used in studies ranging from basic biology to cancer research. It was particularly important in early studies of nutritional requirements, and, recently, it has contributed to our understanding of cell division, the function of mitochondria, glyconeogenesis, the structure of cilia, and ciliate genetics. Only a few other organisms have been studied so intensively.

A primary reason for the success of *Tetrahymena* as an experimental subject is that it is a eukaryote which can be grown in large quantities on bacteria-free media or even on defined media. Further, its cell division can be easily sychronized and its outer membrane readily disrupted for cell-free studies.

The final chapter is flavored, to some extent, by my speculations and opinions, which are rarely found elsewhere in the book. I contend that the documented information available on *Tetrahymena* is sufficient to stimulate even more interest in the organism and that extensive speculation by myself or by others is not always productive.

I would like to express my appreciation to Dr. Ivan L. Cameron and Mr. Glenn Williams, who provided a number of electron micrographs, and to Dr. George W. Kidder and Dr. Virginia C. Dewey, who read an early draft of this manuscript and encouraged me to complete it. My thanks are also due to Mrs. Maria Stone for typing the manuscript and Miss Suzanne Straight for help in proofreading.

DONALD L. HILL

CHAPTER 1

Introduction

Introduction

The phylum Protozoa is comprised of a large group of diverse organisms. Some protozoans have an animal-like nutrition in that they ingest particulate food; others may derive their energy from photochemical reactions, as plants do. Still others, the parasitic protozoans, are able to absorb only dissolved food. The various types of nutrition found in this phylum are accompanied by gross morphological differences and differences in the methods of reproduction. Owing to the widely varying properties found in the phylum, the ciliate *Tetrahymena* is certainly not representative of all Protozoa but may be considered typical of a large group. Its nutrition, morphology, and reproduction are similar to a number of other organisms of this phylum. At least, it may be considered representative of the 6000 species of ciliates that have been described. *Tetrahymena pyriformis* has been studied far more than any other ciliate, and the investigations of its biochemical properties give insight into the relationship of this organism and its relatives to other living things.

Taxonomy

The name *Tetrahymena* is of relatively recent origin; Furgason (1940) coined the word in a classic paper on the buccal ciliature of a group of Protozoa. This name, which has met with almost universal acceptance, is antedated by several other names of allegedly congeneric forms, such as *Leucophra, Acomia, Lambornella, Leucophrydium, Protobalantidium, Ptyxidium, Tetrahymen, Leucophrys, Paraglaucoma, Leptoglena, Trichoda,* and *Turchiniella* (Corliss, 1961). *Tetrahymena* species have occasionally been mistakenly identified as belonging to the genera *Colpidium* and *Glaucoma.* The species name *pyriformis* is quite old. It first appeared as the name for an organism described by Ehrenberg (1830) as *Leucophrys pyriformis.* Later, Maupas (1883, 1888) gave precise and detailed descriptions of two species, *Glaucoma pyriformis* and *Leucophrys patula.* Schewiakoff (1889) also studied this group of organisms and presented his interpretation concering the number and arrangement of buccal ciliary organelles, which became of great significance in the taxonomy of ciliates. In a monumental work on ciliated protozoans, Butschli (1888) adopted Maupas and Stein's (1867) conclusions regarding the taxonomy and morphology of this group. *Tetrahymena pyriformis* has also been called *Tetrahymena geleii.*

Corliss (1961) has put forward the following systematic nomenclature for *Tetrahymena.*

Phylum	Protozoa
Subphylum	Ciliophora
Class	Ciliata
Subclass	Holotricha
Order	Hymenostomatida
Suborder	Tetrahymenina
Family	Tetrahymenidae
Genus	*Tetrahymena*
Species	*Pyriformis, patula, vorax,* etc.

Honigberg *et al.* (1964), a committee set up to review the taxonomy of Protozoa, agree with this classification.

The species included in the genus *Tetrahymena* can be grouped into three complexes. In the *pyriformis* complex are placed *T. pyriformis, T. setifera,* and *T. chironomi.* In the *rostrata* group are assigned *T. rostrata, T. limacis, T. corlissi,* and *T. stegomyiae.* In the *patula* complex are found *T. patula, T. vorax,* and *T. paravorax.* Three doubtful forms are *T. faurei, T. glaucomaeformis,* and *T. parasitica* (Corliss, 1970). A recently discovered species, *T. bergeri* (Roque *et al.,* 1970), has not been assigned to these complexes.

Occurrence

Tetrahymena pyriformis is widely distributed. In fact, Antony van Leeuwenhoek (1676) may have observed this species, or one similar to it, in his earliest descriptions of microorganisms. The habitat of *T. pyriformis* is fresh water, ranging from springs, ditches, creeks, rivers, ponds, and lakes to thermal springs (Elliott, 1970; Elliott and Hayes, 1955; Phelps, 1961). However, Fauré-Fremiet (1912) reported the finding of an organism, probably *T. pyriformis*, in salt marshes; and Sandon (1927) found a similar organism in soil samples. This species apparently enjoys world-wide distribution. It resides in at least forty-six states of the United States and in Canada, Mexico, Panama, Columbia, several countries of Europe, several Pacific Islands, and Australia (Gruchy, 1955; Elliott, 1970; Elliott and Hayes, 1955; Elliott *et al.*, 1962a, 1964).

Although most species of *Tetrahymena* are free-living, some thrive in the bodies of Metazoa. *Tetrahymena limacis* and *Tetrahymena rostrata* parasitize slugs and snails (Kozloff, 1956; Windsor, 1960; Brooks, 1968); whereas *Tetrahymena chironomi* exists in the larvae of midge flies (Corliss, 1960). When some parasitic species are established in culture, they undergo a change involving the loss of some of the cilary meridians (Windsor, 1960; Kozloff, 1962). However, these "lost" meridians return when the organism is reestablished in its host.

Morphology

The body of *Tetrahymena pyriformis* (Figs 1.1 and 1.2) is typically pear-shaped, a characteristic from which the species name is derived. The posterior end is rounded, and the anterior end bluntly pointed. Various modifications, such as ovoid, cucumber, rod, cylindrical, boomerang, and banana forms, are seen occasionally, particularly in old or crowded cultures. The pliability of the pellicle allows for temporary distortion. The mean length times width is about 50 × 30 μ, but twofold variations from these values are not uncommon.

Sixteen to twenty-six rows of cilia (kineties) cover most of the cell surface and lie largely parallel to the long axis of the cell. The extreme anterior and posterior ends are bare. The buccal or oral apparatus lies toward the anterior end and consists of a shallow cavity with a membrane of fused cilia, called the "undulating membrane," on the right side and three smaller membranes lying parallel on the left wall of the cavity. On ingestion of food, food vacuoles are formed from the buccal cavity; and fibrils may be seen marking the site of formation. The anal pore (cytoproct, cytopyge) lies on the posterior, ventral body surface; and, typically, two contractile vacuole pores lie on the pellicle at the right posterior side. Located in the

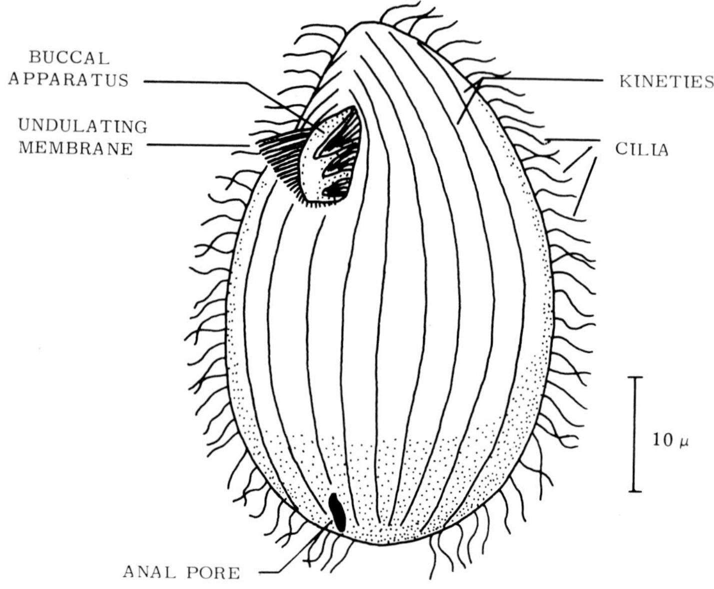

FIG. 1.1. External features of *Tetrahymena*.

central portion of the body, a single, round macronucleus about 10 μ in diameter is present. In sexually active strains, there is a micronucleus, and a Golgi apparatus is visible in these strains when they are starved for induction of conjugation. The mitochondria and endoplasmic reticulum resemble those of other organisms.

Much of the above descriptive work was done by Furgason (1940), Corliss (1953a), Roth and Minick (1961), Elliott *et al.* (1962b), and Elliott and Zieg (1968).

More detailed morphological studies have been made on the pellicle, the buccal apparatus, the nuclei, and the expulsion vesicle of *T. pyriformis*. Studies on the pellicle quite often are made after impregnation with silver, a procedure which allows much better visualization of the cortical structures. (Corliss, 1953b; Pitelka, 1961; Frankel and Heckmann, 1968). By use of this technique, the development of cortical patterns on the cell has been

---

FIG. 1.2. Midlongitudinal section of *Tetrahymena pyriformis* HSM. Abbreviations: UM, cross section of undulating membrane; BM, buccal membranelle; BB, basal bodies; CPP, cytopharyngeal pouch; MIT, mitochondria; CB, chromatin bodies; NO, nucleoli; MA, macronucleus; MI, micronucleus; MU, mucocysts; EVP, expulsion vesicle pore; FV, food vacuole (empty). Magnification: ×7500. (Courtesy of Dr. I. L. Cameron and Mr. Glenn Williams.)

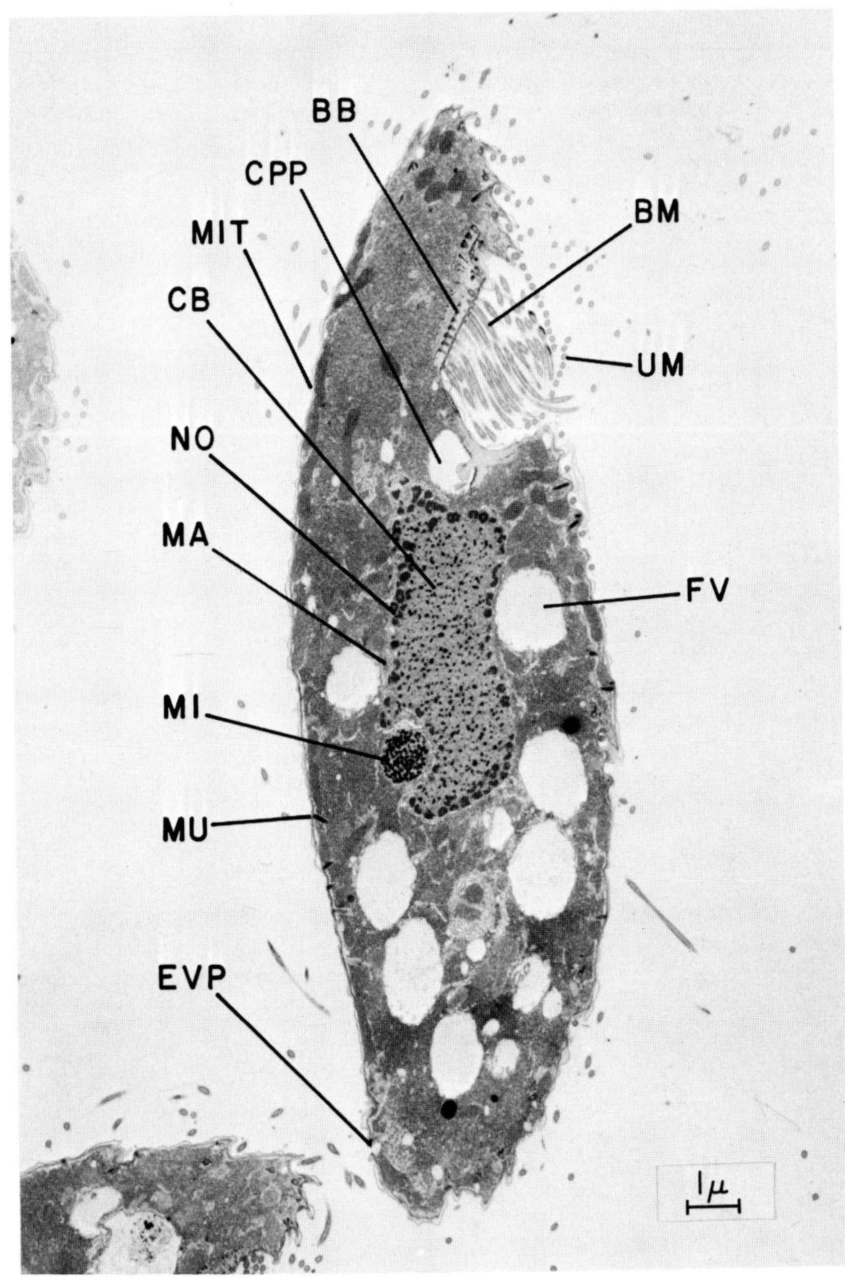

FIG. 1.2.
See facing page for legend.

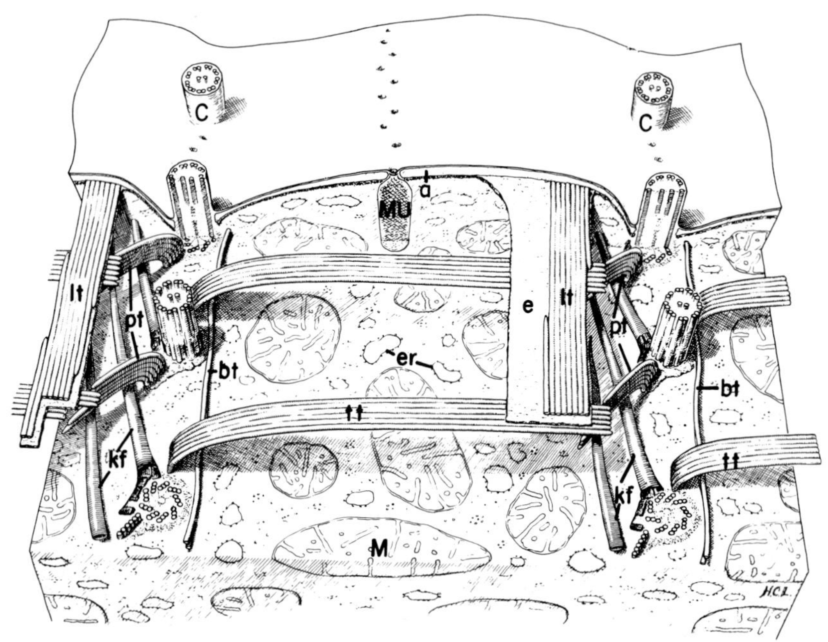

FIG. 1.3. Three-dimensional concept of the structures of the somatic cortex of *Tetrahymena pyriformis*. At the left and right are segments of two kineties with basal portions of their cilia (C). A portion of a secondary meridian lies midway between the primary meridians and contains a cutaway of a mucocyst (MU). A lattice work of three prominent sets of microtubules is present just under the pellicle: the longitudinal set (lt), the transverse set (tt), and the postciliary set (pt). Below this lattice-work is a band of two basal microtubules (bt) running along one side of each kinety at the proximal end of the basal bodies. A triangular support for the basal body–cilium complex is formed by one of the microtubular bands and the kinetodesmal fiber (kf). Mitochondria (M), rough endoplasmic reticulum (er), and ribosomes are depicted in the cytoplasm as well as the amorphous ectoplasmic material (e) and alveoli (a) associated with the pellicle. Magnification: approximately ×40,000. (Drawn by Mrs. H. C. Lyman; provided by Dr. R. D. Allen; from Allen, 1968.)

investigated (Nanney, 1966a,b,c,d; 1967a,b; 1968a,b; 1970; 1971). A number of examined cortical characteristics of *T. pyriformis* have little taxonomic utility because of wide variation within the different syngens (varieties). However, one property, that of the position of the expulsion vesicle pores (contractile vacuole pores) in relation to the first ciliary meridian, may be useful in distinguishing the different syngens. The fine structure and morphogenesis of ciliary basal bodies have been studied with the use of an electron microscope (Metz and Westfall, 1954; R. D.

Allen, 1967, 1969). Each cilium, 5–8 μ in length, is attached to a subsurface kinetosome. The kineties consist of cilia, kinetosomes, and kinetodesmal (longitudinal) fibrils. The fibrils overlap like shingles to form the kinetodesma seen with the light microscope. Figure 1.3 is a diagram of the cortical structures of *T. pyriformis* (Allen, 1968). Cortical membranes

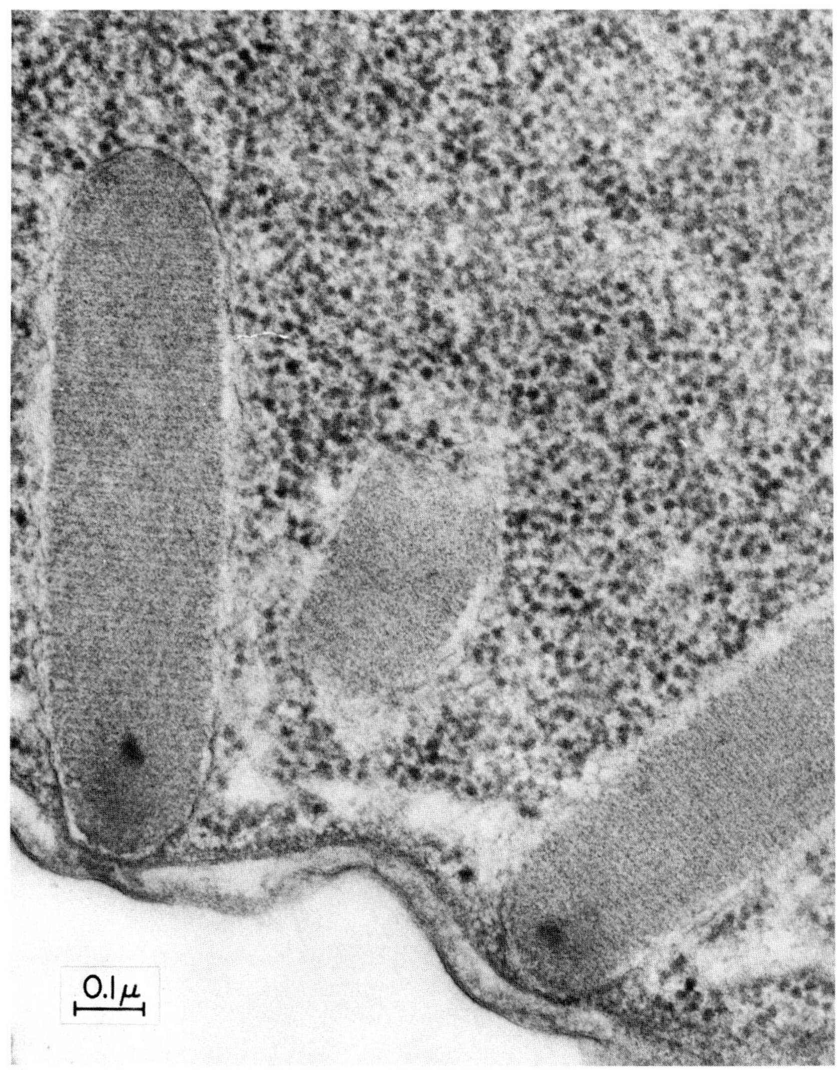

FIG. 1.4. Mucocysts of *Tetrahymena pyriformis* HSM showing crystalline lattice, outer limiting membrane, and undefined density near pellicle. Magnification: ×100,000. (Courtesy of Dr. I. L. Cameron and Mr. Glenn Williams.)

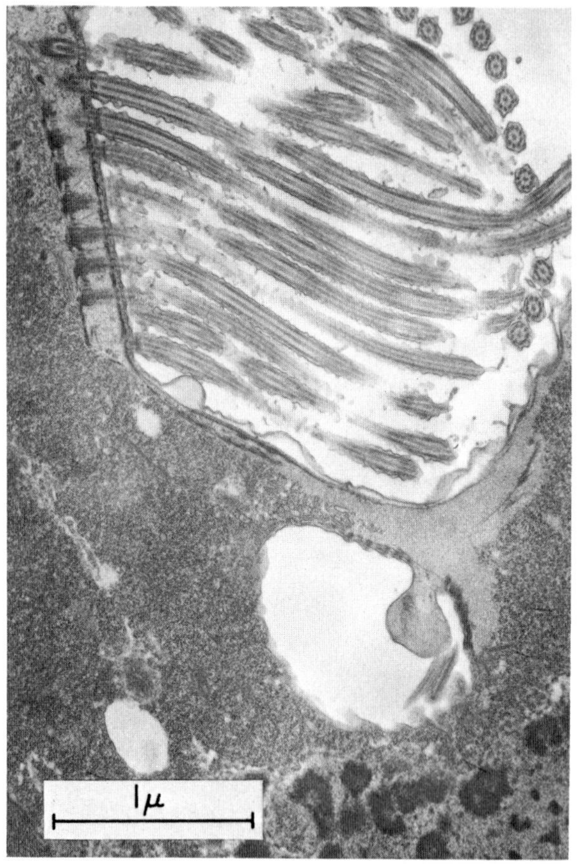

FIG. 1.5. Base of oral groove showing cytopharyngeal pouch. Cilia are arranged to move food into the pouch. Magnification: ×27,000. (Courtesy of Dr. I. L. Cameron and Mr. Glenn Williams.)

and the attached cortical structures can be obtained by homogenization in dilute ethanol and differential centrifugation (Hufnagel, 1968).

One component of ciliary preparations, originally thought to be a part of the cilia, is now identified as mucocysts or protrichocysts (Fig. 1.4) (Alexander, 1968). When the cells are under stress, these structures discharge; and the exuded material produces a mucilaginous sheath over the surface (Tokuyasu and Scherbaum, 1965). Antisera produced against this excretion immobilize the cilia (Alexander, 1968). The mucoysts are obtained intact by treating the cells with calcium chloride and then extracting with dilute ethanol containing a chelating agent. Mucocysts and other structures of *T. pyriformis* can be visualized after fixing the cells with a mixture of glutaraldehyde and tris(1-aziridinyl)phosphine oxide (Williams and Luft, 1968).

Electron-microscopic studies have been made of the oral area of *T. pyriformis* (Fig. 1.5) (Nilsson and Williams, 1966, Wolfe, 1970), *Tetrahymena patula* (Miller and Stone, 1963), and *Tetrahymena vorax* (Buhse, 1968); and considerable effort has been expended in studying the morphogenesis of the buccal apparatus (Williams and Zeuthen, 1966; Frankel, 1967, 1969; R. D. Allen, 1969). Procedures are now available for the isolation of intact oral structures (Williams and Zeuthen, 1964; Whitson *et al.*, 1966).

The fine structure of the macronucleus has also been investigated (Roth and Minick, 1958, 1961; Elliott *et al.*, 1962b; Swift *et al.*, 1964; Cameron *et al.*, 1966; Flickinger, 1965; Falk *et al.*, 1968; Tamura *et al.*, 1969; Satir and Dirksen, 1971). This organelle is surrounded by a double-layered envelope (Roth and Minick, 1961), and fragments of this envelope can be isolated (Franke, 1967). The macronuclear envelope contains numerous pores (Fig. 1.6). There are 80–190 pores per square micron, and the pores are

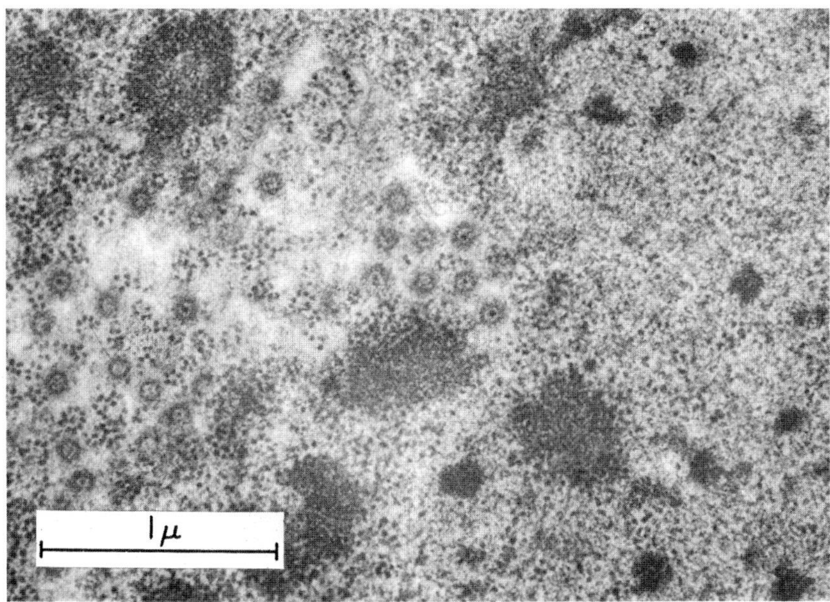

FIG. 1.6. Tangential section of nuclear envelope showing nuclear pores containing central density. Polyribosomes are visible in the cytoplasm. Nuclear material is on the right. Magnification: ×32,000. (Courtesy of Dr. I. L. Cameron and Mr. Glenn Williams.)

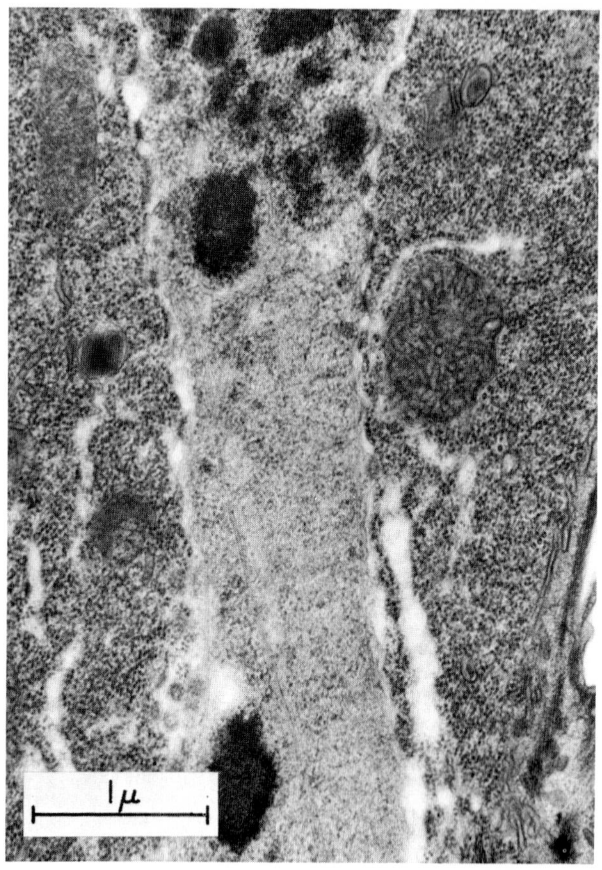

Fig. 1.7. Portion of dividing macronuclei showing microtubules (mt). Magnification: ×24,000. (Courtesy of Dr. I. L. Cameron and Mr. Glenn Williams.)

from 45 to 75 nm in diameter (Wunderlich, 1968, 1969). About 30% of the nuclear surface is pores. The pore diameter varies with the method of preparation of the nuclear envelope for electron microscopy (Speth and Wunderlich, 1970). The nucleoplasm contains chromatin granules and 200–1000 peripheral, larger particles identified as nucleoli (Roth and Minick, 1958, 1961; Swift *et al.*, 1964; Charret, 1969; Cameron and Guile, 1965; Abdel-Hameed, 1969; Gorovsky, 1965; Nilsson, 1969; Satir and Dirksen, 1971). Chromatin fibrils, 4–10 nm in diameter, can be detected by electron microscopy (Sato and Saito, 1959; Abdel-Hameed, 1969; Lee and Byfield, 1969). During division, intramacronuclear microtubules of the spindle type can be found (Fig. 1.7); they probably participate in the process of division

(Falk *et al.*, 1968; Ito *et al.*, 1968) and particularly in the process of nuclear elongation prior to division (Tamura *et al.*, 1969). Procedures for isolating the macronuclei from *T. pyriformis* have been developed (Lee and Scherbaum, 1965; Mita *et al.*, 1966; Prescott *et al.*, 1966; Gorovsky, 1970).

The micronucleus (Fig. 1.8), when present, lies adjacent to the macronucleus, is 1–2 μ in diameter, and contains dense, Feulgen-positive material. It also is surrounded by a double membrane which contains pores (Gorovsky, 1970). Recently developed techniques allow micronuclei to be isolated (Muramatsu, 1970; Gorovsky, 1970). On division of the cell, a diploid set of five chromosomes arise from this structure (Ray, 1956a). A procedure is

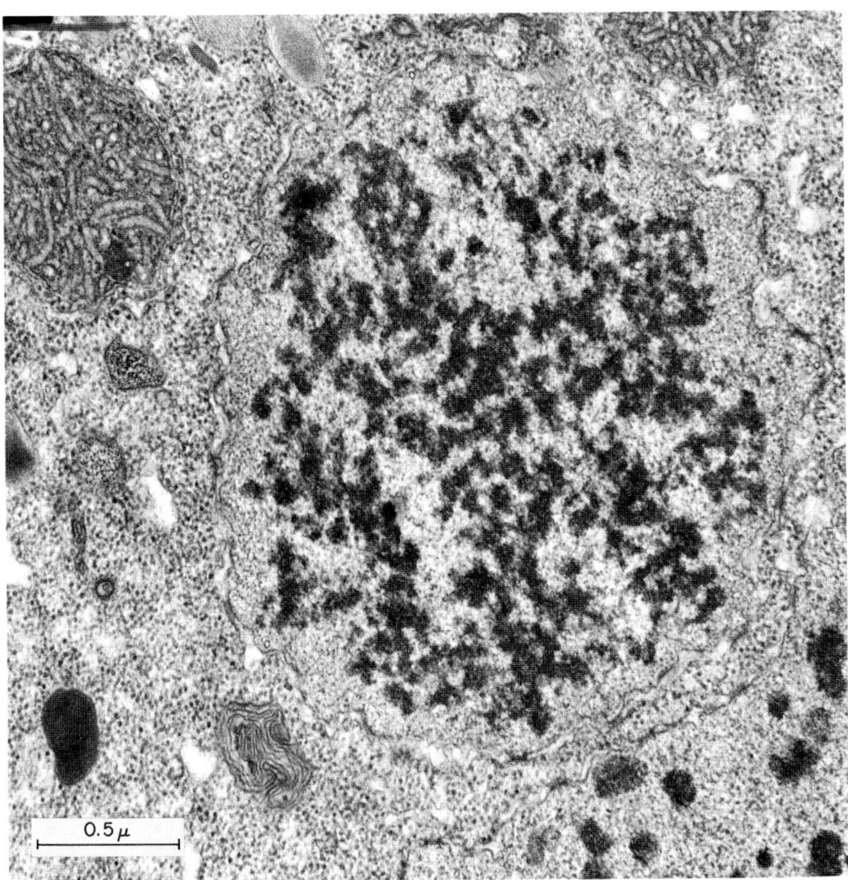

FIG. 1.8. Micronucleus showing condensed chromatin and micronuclear envelope. Structure is adjacent to macronucleus. Magnification: ×65,000. (Courtesy of Dr. I. L. Cameron and Mr. Glenn Williams.)

available for preparing these for visualization by photomicrography (Ray, 1956b).

The explusion vesicles (Fig. 1.9) contribute to the maintenance of constant cell volume by eliminating fluid at a rate equal to the passive entry of water into the cell. The vesicles are surrounded by a single membrane with a thickness of 7 nm (Elliott and Bak, 1964). Early in the process of evacuation, water in the vesicle flows back into the collecting tubules. Later, the vesicle moves toward one of the outlet pores (Cameron and Burton, 1969). The expulsion vesicle pores indent immediately preceding

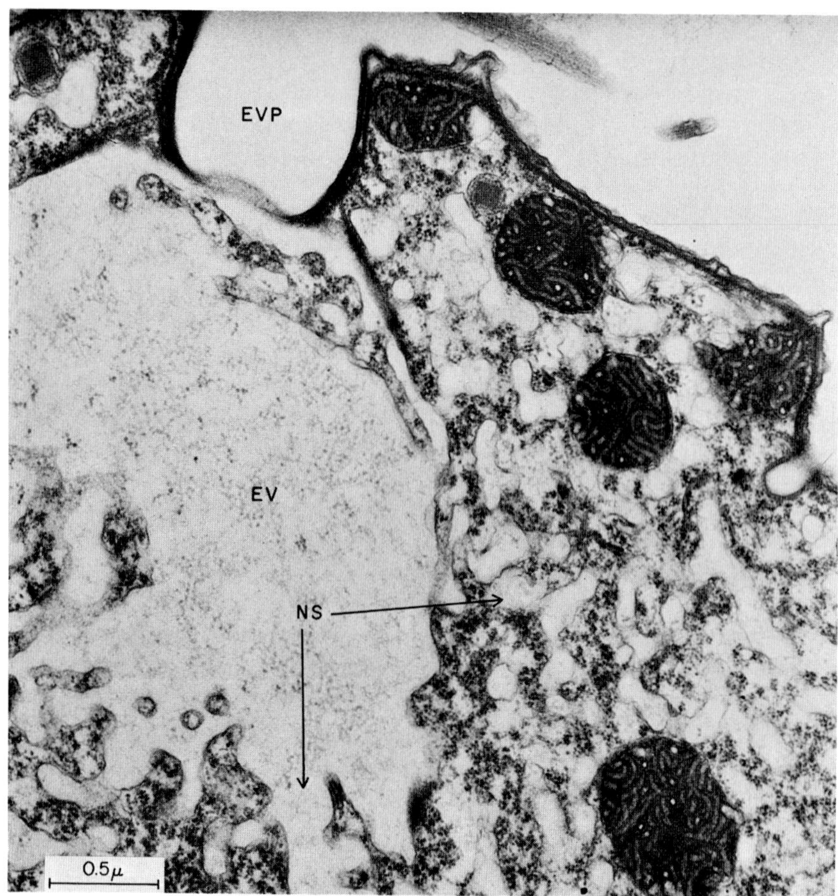

FIG. 1.9. Expulsion vesicle and port showing nephridial system. Abbreviations: EVP, expulsion vesicle pore; EV, expulsion vesicle; NS, nephridial system. Magnification: ×50,000. (Courtesy of Dr. I. L. Cameron and Mr. Glenn Williams.)

the expulsion of the vesicular fluid (Dunham and Stoner, 1969). To expel water, the vesicle itself does not contract; instead, the cytoplasm pushes against it, and the water passes through one of the two outlet pores to the outside of the cell (Organ *et al.*, 1968; 1969; Cameron and Burton, 1969).

As is the case for many other organisms, *T. pyriformis* can regenerate large portions of its body following amputation (Albach, 1959). Both anterior and posterior sections give rise to intact cells; but large, anterior fragments regenerate much better than small, posterior fragments. Some of the visible morphological structures can be regenerated in their entirety.

Ingestion and Digestion

Elliott and Clemmons (1966) have made a detailed, ultrastructural study of ingestion and digestion in *Tetrahymena pyriformis*. Food vacuoles, which are formed at the base of the buccal apparatus, fuse with pinocytotic vacuoles, derived from the surface membrane, and with lysosomes, probably derived from the endoplasmic reticulum (Elliott, 1965), thereby acquiring the hydrolases present in these structures (Elliott and Clemmons, 1966; Müller *et al.*, 1963). Two types of lysosomes are present. One type, with a density of 1.24 is rich in α-glucosidase, phosphatase, deoxyribonuclease, α-amylase, and β-*N*-acetylglucosaminidase. The other type, with a density of 1.14 contains predominantly ribonuclease and proteinase (Müller *et al.*, 1966; Müller, 1970). When the cells are maintained in a dilute salt solution, the high-density particles disappear and their enzymes appear in the medium.

Digestion proceeds to completion in the food vacuole, and undigested materials are lost through the anal pore. *Tetrahymena corlissi* has a similar food vacuole cycle (Müller and Rohlich, 1961). In stationary growth, *T. pyriformis* develops autophagic vacuoles which contain mitochondria and other cellular particles (Elliott and Clemmons, 1966), and the number of lysosomes decreases (Elliott *et al.*, 1966).

By comparing mathematically the rates of food vacuole formation with rates of acetate and glucose utilization in proteose–peptone medium, Seaman (1961, 1962) concluded that uptake through the mouth region is very slow and that most of the material utilized must come through the cell membrane. Seaman (1961) also reported that ingestion of fluid by vacuole formation requires an inducer provided in the proteose–peptone medium and that cells grown in a defined medium cannot ingest fluid. Some attempts were made to purify this factor (Seaman and Mancilla, 1961). However, extensive investigations on vacuole formation refute Seaman's claim that vacuoles are not present in cells grown in defined media

(Nilsson, 1968; Nilsson and Chapman-Andresen, 1968; Chapman-Andresen and Nilsson, 1968). These studies show that log-phase cells contain up to eight vacuoles and that the number increases on starvation. The increase in food vacuole formation stops when nutrients are added, and the structures then dissolve. Vacuole formation also stops during division and in the presence of 2,4-dinitrophenol.

Pattern Formation

When in a dense culture and in a shallow medium, *Tetrahymena pyriformis* swims en masse in streams which form patterns (Fig. 1.10) (Loefer and Mefferd, 1952; Jahn and Brown, 1961; Jahn *et al.*, 1961). The streams are horizontal and form a polygonal network, with four or five streams meeting at each node of the net. At the nodes, the organisms fall toward the bottom, but then swim upward to rejoin the horizontal streams. Under optimal conditions, the patterns can be formed in less than 10 sec.

Factors that influence the motility of cells also affect formation of the pattern (Loefer and Mefferd, 1952). Age is very important, since young cultures form more distinct patterns and form them more quickly. Cells dividing synchronously have a minimum rate of pattern formation shortly

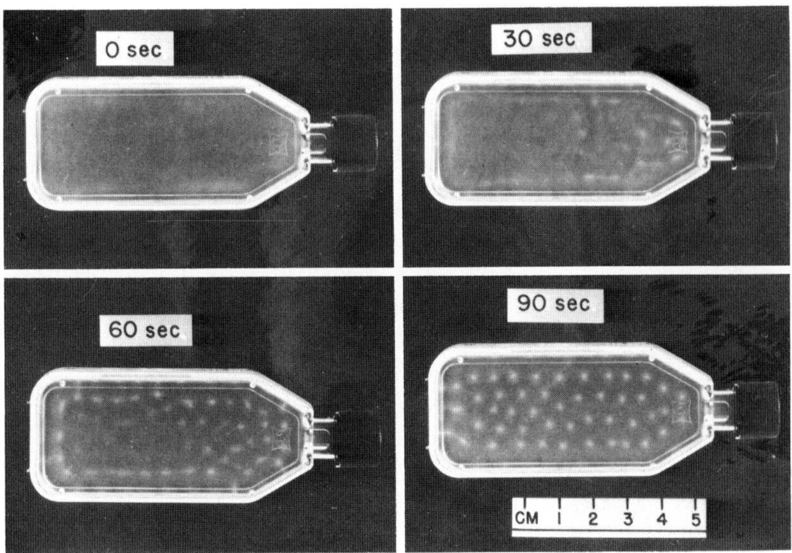

Fig. 1.10. Pattern formation by *Tetrahymena pyriformis*. (Provided by Dr. J. J. Wille, Jr.; from Wille and Ehret, 1968b.)

after division (Wille and Ehret, 1968b). Aggregation of organisms can be stimulated to develop by phenanthraquinone (Jones and Baker, 1946).

Cell Size

The average cell size, normally 50 × 30 μ, varies with a number of conditions. An important factor is the proteose–peptone concentration of the medium. Cell size increases with proteose–peptone concentration up to 4%, and this increase is independent of the pH of the medium (Huddleston *et al.*, 1964). At concentrations above 4%, osmolarity has a strong effect, and the cells shrink. Sucrose (0.1 *M*) has a similar effect (Stoner and Dunham, 1970). Cell size is reduced by 50% during eight generations of logarithmic growth. In the course of this growth, cells are normally distributed with respect to size; but as the culture approaches stationary phase, the population splits into groups of large and small cells (Scherbaum, 1956). A notable decrease in cell size can be attained by transferring cells from logarithmically growing cultures to a medium devoid of nutrients (Cameron, 1967). Cells divide two or more times in such a medium and become much smaller.

Cell size is related to the temperature at which the culture is maintained. It decreases with increasing temperature in the range of 10° to 30°C (James and Read, 1957; Thormar, 1962a). At temperatures above optimum, the cells enlarge again. During heat-shock treatment, the size increases up to 300% but returns to normal during the subsequent synchronous divisions (Scherbaum, 1956; Schmid, 1967b).

Volume growth curves for individual cells follow linear courses throughout interphase (Cameron and Prescott, 1961; Summers, 1963). Such linearity is interpreted to mean that the growth from division to division is not autocatalytic; that is, growth does not lead to an increased rate of growth (Cameron and Prescott, 1961). One report states that the volume growth rate drops to zero during a small portion of the cell cycle (Summers, 1963); another asserts that the volume increases much more rapidly during division, an increase tentatively attributed to an elevated rate of water uptake (Cameron and Prescott, 1961).

Growth Conditions

Tetrahymena can grow in a medium which is defined (Kidder and Dewey, 1951) (Table 1.1), which is undefined but axenic (Lwoff, 1923), or which contains bacteria as a food source. However, media that are not free of bacteria are of little use to the biochemical protozoologist. In many cases,

TABLE 1.1

COMPOSITION OF BASAL MEDIUM[a]

Substance	(μg/ml)	Substance	(μg/ml)
DL-Alanine	110	Ca pantothenate	0.10
L-Arginine	206	Niacin	0.10
L-Aspartic acid	122	Pyridoxal·HCl	0.10
Glycine	10	Pyridoxamine·HCl	0.10
L-Glutamic acid	233	Riboflavin	0.10
L-Histidine	87	Folic acid	0.01
DL-Isoleucine	276	Thiamine·HCl	1.00
L-Leucine	344	Biotin	0.0005
L-Lysine	272	Choline·Cl	1.00
DL-Methionine	248	DL-Lipoic acid	0.004
L-Phenylalanine	160		
L-Proline	250	$MgSO_4 \cdot 7\ H_2O$	100.00
DL-Serine	394	$Fe(NH_4)_2(SO_4)_2 \cdot 6\ H_2O$	25.00
DL-Threonine	326	$MnCl_2 \cdot 4\ H_2O$	0.50
L-Tryptophan	72	$ZnCl_2$	0.05
DL-Valine	162	$CaCl_2 \cdot 2\ H_2O$	50.00
		$CuCl_2 \cdot 2\ H_2O$	5.00
Uracil	10	$FeCl_3 \cdot 6\ H_2O$	1.25
Adenylic acid	20	K_2HPO_4	1000.00
Cytidylic acid	25	KH_2PO_4	1000.00
Guanylic acid	30		
Na acetate	1000	Tween 80[c]	10,000.00
Glucose[b]	2500		

[a] From Kidder and Dewey (1957). The components of this medium which are not essential serve to stimulate growth.

[b] Autoclaved separately and added separately.

[c] Optional, see Chapter 3.

a defined medium is a necessity. Growth requirements in the defined medium for *T. pyriformis* are given in Table 1.2. Since these components are commonly found in biochemical laboratories, this protozoan can be grown without any great difficulty. A further asset in using the defined medium is that quantitation of results is quite easy; and the high growth rate assures a generous yield of organisms. A special centrifuge has been developed for harvesting mass cultures of *T. pyriformis* (Conner *et al.*, 1966).

A recent, important observation is that particulate material is required for rapid cell multiplication (Rasmussen and Kludt, 1970). Growth in sterile-filtered proteose–peptone is improved by addition of insoluble materials such as talcum, quartz, clay, and calcium carbonate and by addition of water-soluble compounds that produce insoluble particles in aqueous solution. The authors suggest that formation of food vacuoles, induced by the presence of particulate material, is a condition for rapid growth. Further, they surmise that some of the essential nutrients present in proteose–peptone must enter the cells via food vacuoles.

Over a wide range, pH has little effect on growth (Wingo and Anderson, 1951). The organism, of course, will not grow in very acidic (below pH 5.0) or very basic media (above pH 8.6) (Prescott, 1958).

Tetrahymena pyriformis is able to adjust somewhat to heat or to cold, but such adjusted cells are more sensitive to the opposite extreme (Thormar, 1962b). Heat-adapted strains can grow at 40°C (Holz *et al.*, 1959), and a particular strain isolated from a hot spring is able to multiply at 41.2°C (Phelps, 1961).

The length of the lag period when cells are placed in a fresh medium is independent of the size of the inoculum but is dependent upon and proportional to the age of the inoculum (Phelps, 1935; Prescott, 1957a).

The generation time for *T. pyriformis* varies depending upon the strain used and the conditions of growth; but it is generally about 3 or 4 hr (Prescott, 1957a,b; Cameron and Prescott, 1961; Szyszko *et al.*, 1968). The temperature at which maximum growth is normally achieved is about 29°C (Phelps, 1946; Thormar, 1962b; Mackenzie *et al.*, 1966), but this also varies with the strain and the composition of the medium. The effects of temperature on the generation time and the different parts of the cell cycle are known (Prescott, 1957b; Mackenzie *et al.*, 1966). Stages G_1, S, and G_2

TABLE 1.2

SUBSTANCES NECESSARY IN DEFINED MEDIUM FOR GROWTH OF *Tetrahymena pyriformis*

Amino acids	Vitamins	Inorganic	Other
Arginine	Thiamine	Phosphate	Guanine
Histidine	Riboflavin	Magnesium	Uracil or cytidine or cytidylic acid
Isoleucine	Pyridoxal	Potassium	
Leucine	Pantothenate	Copper	
Lysine	Folate		
Methionine	Niacin		
Phenylalanine	Lipoic (thioctic) acid		
Threonine			
Tryptophan			
Valine			

all decrease as the temperature increases from 17° to 26°C. Not much change occurs as the temperature rises 3° more, but with a 6° rise, all three portions of the cell cycle increase. Cells in the exponential growth phase (ultradian) are strongly temperature-dependent, but those in a non-stationary growth phase (infradian) show little or no temperature dependence between 15° and 27°C (Szyszko *et al.*, 1968).

Growth is stopped below 10°C, but in 10% dimethyl sulfoxide, cells can survive quick-freezing to very low temperatures. One procedure calls for reducing the temperature to −20°C in two steps, then quickly to −196°C (Hwang *et al.*, 1964). Another method uses a drop of 4.5°/min to freezing, and then the temperature is rapidly dropped to −196°C (Wang and Marquardt, 1966).

The heat shocks used to induce synchrony have a marked effect on the morphology of the cells undergoing treatment. Formation of oral structures is stopped, and cells with partially formed oral structures resorb them during the heat treatment (Frankel, 1962; 1964a,b,c; Holz, 1960; Holz *et al.*, 1957; Williams and Scherbaum, 1959; Williams *et al.*, 1960; Williams, 1964b; Gavin, 1965). After a certain degree of completion, however, the oral primordium becomes stabilized and is no longer resorbable (Williams, 1964b; Frankel, 1964c). Supraoptimal temperatures induce lesions in other types of cell organelles (Levy *et al.*, 1969). These changes include loss of tubular structures from mitochondria, disappearance of rough endoplasmic reticulum, nuclear abnormalities, and a swelling of bodies which are either peroxisomes or primary lysosomes. Oddly, the heated cells sometimes do not complete division, with the result that chains are formed (Frankel, 1964b). After termination of the heat shocks, however, the chains break up and give rise to normal cells.

The addition of kinetin (6-furfurylaminopurine) to the medium following heat shocks results in increased proportions of synchronously dividing cells (Ron and Guttman, 1961). Not only is the morphology of the cell affected by heat shocks, but many biochemical parameters are changed. These alterations are not discussed here, but in the appropriate chapter.

Synchronization

For most strains, a division block is permanent at 33.9°C (Schmid, 1967a; Scherbaum and Zeuthen, 1954, 1955; Scherbaum, 1956), and this blockade is frequently utilized to induce synchronous division (Scherbaum and Loefer, 1964; Williams, 1964a). However, strain WH6 requires a higher shocking temperature, 42.8°C, lethal for other strains (Holz *et al.*, 1957). For individual cells, there is increasing sensitivity to heat shocks with increasing age up to a sudden "transition point," at which loss of sensitivity occurs (Williams, 1964b). There is a sharp increase in sensitivity just before the transition point. Speculation is that the transition point may occur at a certain stage of development of the oral structure, and dedifferentiation of the partially formed structure is thought to be involved (Williams, 1964b; Nachtwey, 1961).

Other methods that have been used to synchronize *Tetrahymena pyriformis* are a sudden change in pH (Prescott, 1958), cold shocks (Zeuthen and Scherbaum, 1954; Scherbaum and Loefer, 1964; Padilla and Cameron, 1964; Debault and Ringertz, 1967), centrifugation (Corbett, 1964), temporary limitation of access to thymidine compounds (Villadsen and Zeuthen, 1969), hypoxic shocks (Rasmussen, 1963; Rooney and Eiler, 1967, 1969), sudden exposure to visible light (Wille and Ehret, 1968a), starvation and refeeding (Cameron and Jeter, 1970), exposure to colchicine or Colcemid (Wunderlich and Peyk, 1969), and reversible inhibition with vinblastine (Stone, 1968a,b). Another species of *Tetrahymena, T. vorax,* can also be synchronized; but the degree of synchrony is not as great (Williams, 1964a). Several kinds of apparatus have been developed for induction of growth oscillations in Protozoa. Such equipment is designed to regulate automatically the variables involved (Scherbaum and Zeuthen, 1955; Lee, 1959; James, 1961; Scherbaum and Jahn, 1964). Dense populations of cells for synchronous division, and for other purposes, can be attained by cultivating them in rotating glass bottles (Hjelm, 1970).

Literature

The number of papers published on *Tetrahymena* increased markedly after the organism was established in axenic culture (Lwoff, 1923), and there was yet another increase after a defined medium was developed for the organism (Kidder and Dewey, 1951). Kidder and Dewey (1951), in addition to reporting the composition of the defined medium, also reviewed the nutritional requirements and the metabolism of *Tetrahymena* as they were known at that time. In 1954, Corliss took note of the explosion of interest in this protozoan in an article considering the increased curiosity about the ciliate. Since 1954 the number of articles published each year on *Tetrahymena* has continued to increase; and such an expansion of interest has led to the publication of a number of reviews concerning this genus. Seaman (1955) reviewed knowledge of the cellular constituents and metabolism but also engaged in much speculation which has proved to be of little value. Elliott (1959a) briefly considered the entire biology of the organism and also (Elliott, 1959b) reminisced over a quarter century of his investigations on *Tetrahymena*. Seaman and Reifel (1963) considered, this time briefly, the chemical composition and metabolism of *Tetrahymena* in a review which spanned the entire phylum. Recently, there have appeared a flood of reviews, each considering separate aspects of the biology of Protozoa. As a summary, the following subjects have been discussed: genetics (Kimball, 1964; Allen, S. L. 1967, 1968; Preer, 1969); nutrition (Holz, 1964; Hall, 1967; Kidder, 1967); synchrony (Scherbaum, 1960; Scherbaum and Loefer, 1964); respiration (Danforth, 1967); morphogenesis (Tartar, 1967); nuclear replication (Prescott and Stone, 1967); effects of radiation (Giese, 1967); carbohydrates (Ryley, 1967); nitrogen metabolism (Kidder, 1967); lipid composition and metabolism (Dewey, 1967); transport (Conner, 1967); digestion (Müller, 1967); cilia and other fibrillar systems (Child, 1967; Pitelka, 1969); development (Hanson, 1967); and nucleic acids (Mandel, 1967). In nearly all of these reviews, *Tetrahymena* is not the only organism considered; in some cases, it is discussed only briefly.

REFERENCES

Abdel-Hameed, F. (1969). *J. Protozool.* **16,** Suppl. 19.

Albach, R. A. (1959). *Trans. Amer. Microsc. Soc.* **78,** 276.

Alexander, J. B. (1968). *Exp. Cell Res.* **49,** 425.

Allen, R. D. (1967). *J. Protozool.* **14,** 553.

Allen, R. D. (1969). *J. Cell Biol.* **40,** 716.

Allen, S. L. (1967). *In* "Chemical Zoology" (M. Florkin and B. T. Scheer, eds.), Vol. 1 (G. W. Kidder, ed.), p. 617. Academic Press, New York.

Allen, S. L. (1968). *Ann. N. Y. Acad. Sci.* **151,** 190.

Brooks, W. M. (1968). *Hilgardia* **39,** 205.

Buhse, H. E., Jr. (1968). *J. Protozool.* **15,** Suppl. 10.

Butschli, O. (1888). *In* "Klassen u. Ordnung d. Thier-Reichs" (H. G. Bronn, ed.), p. 1098. Leipzig, 1 (III abt.).

Cameron, I. L. (1967). *J. Protozool.* **14,** Suppl. 7.

Cameron, I. L., and Burton, A. L. (1969). *Trans. Amer. Microsc. Soc.* **88,** 386.

Cameron, I. L., and Guile, E. E. (1965). *J. Cell Biol.* **26,** 845.

Cameron, I. L., and Jeter, J. R., Jr. (1970). *J. Protozool.* **17,** 429.

Cameron, I. L., and Prescott, D. M. (1961). *Exp. Cell Res.* **23,** 354.

Cameron, I. L., Padilla, G. M., and Miller, O. L., Jr. (1966). *J. Protozool.* **13,** 336.

Chapman-Andresen, C., and Nilsson, J. R. (1968). *C. R. Trav. Lab. Carlsberg* **36,** 405.

Charret, R. (1969). *Exp. Cell Res.* **54,** 353.

Child, F. M. (1967). *In* "Chemical Zoology" (M. Florkin and B. T. Scheer, eds.), Vol. 1 (G. W. Kidder, ed.), p. 381. Academic Press, New York.

Conner, R. L. (1967). *In* "Chemical Zoology" (M. Florkin and B. T. Scheer, eds.). Vol. 1 (G. W. Kidder, ed.), p. 309. Academic Press, New York.

Conner, R. L., Cline, S. G., Koroly, M. J., and Hamilton, B. (1966). *J. Protozool.* **13,** 377.

Corbett, J. J. (1964). *Exp. Cell Res.* **33,** 155.

Corliss, J. O. (1953a). *Parasitology* **43,** 49.

Corliss, J. O. (1953b). *Stain Technol.* **28,** 97.

Corliss, J. O. (1954). *J. Protozool.* **1,** 156.

Corliss, J. O. (1960). *Parasitology* **50,** 111.
Corliss, J. O. (1961). "The Ciliated Protozoa: Characterization, Classification, and Guide to the Literature." Pergamon Press, New York.
Corliss, J. O. (1970). *J. Protozool.* **17,** 198.
Danforth, W. F. (1967). *In* "Research in Protozoology" (T-T. Chen, ed.), Vol. 1, p. 201. Pergamon Press, Oxford.
Debault, L. E., and Ringertz, N. R. (1967). *Exp. Cell Res.* **45,** 509.
Dewey, V. C. (1967). *In* "Chemical Zoology" (M. Florkin and B. T. Scheer, eds.), Vol. 1 (G. W. Kidder, ed.), p. 162. Academic Press, New York.
Dunham, P. B., and Stoner, L. C. (1969). *J. Cell Biol.* **43,** 184.
Ehrenberg, C. G. (1830). *Abh. Deut. Akad. Wiss. Berlin* (gedruckt 1932) 1.
Elliott, A. M. (1959a). *Annu. Rev. Microbiol.* **13,** 79.
Elliott, A. M. (1959b). *J. Protozool.* **6,** 1.
Elliott, A. M. (1965). *Science* **149,** 640.
Elliott, A. M. (1970). *J. Protozool.* **17,** 162.
Elliott, A. M., and Bak, I. J. (1964). *J. Protozool.* **11,** 250.
Elliott, A. M., and Clemmons, G. L. (1966). *J. Protozool.* **13,** 311.
Elliott, A. M., and Hayes, R. E. (1955). *J. Protozool.* **2,** 75.
Elliott, A. M., and Zieg, R. G. (1968). *J. Cell Biol.* **36,** 391.
Elliott, A. M., Addison, M. A., and Carey, S. E. (1962a). *J. Protozool.* **9,** 135.
Elliott, A. M., Kennedy, J. R., and Bak, I. J. (1962b). *J. Cell Biol.* **12,** 515.
Elliott, A. M., Studier, M. A., and Work, J. A. (1964). *J. Protozool.* **11,** 370.
Elliott, A. M., Travis, D. M., and Work, J. A. (1966). *J. Exp. Zool.* **161,** 177.
Falk, H., Wunderlich, F., and Franke, W. W. (1968). *J. Protozool.* **15,** 776.
Fauré-Fremiet, E. (1912). *Arch. Anat. Microsc. Morphol. Exp.* **13,** 401.
Flickinger, C. J. (1965). *J. Cell Biol.* **27,** 519.
Franke, W. W. (1967). *Z. Zellforsch. Mikrosk. Anat.* **80,** 585.
Frankel, J. (1962). *C. R. Trav. Lab. Carlsberg* **33,** 1.
Frankel, J. (1964a). *Exp. Cell Res.* **35,** 349.
Frankel, J. (1964b). *J. Protozool.* **11,** 514.
Frankel, J. (1964c). *J. Exp. Zool.* **155,** 403.
Frankel, J. (1967). *J. Cell Biol.* **35,** 165A.
Frankel, J. (1969). *J. Protozool.* **16,** 26.
Frankel, J., and Heckmann, K. (1968). *Trans. Amer. Microsc. Soc.* **87,** 317.
Furgason, W. H. (1940). *Arch. Protistenk.* **94,** 224.
Gavin, R. H. (1965). *J. Protozool.* **12,** 307.
Giese, A. C. (1967). *In* "Research in Protozoology (T-T. Chen, ed.), Vol. 2, p. 267. Pergamon Press, Oxford.
Gorovsky, M. A. (1965). *J. Cell Biol.* **27,** 37A.
Gorovsky, M. A. (1970). *J. Cell Biol.* **47,** 619.
Gruchy, D. F. (1955). *J. Protozool.* **2,** 178.
Hall, R. P. (1967). *In* "Research in Protozoology" (T-T. Chen, ed.), Vol. 1, p. 337. Pergamon Press, Oxford.
Hanson, E. D. (1967). *In* "Chemical Zoology" (M. Florkin and B. T. Scheer, eds.), Vol. 1 (G. W. Kidder, ed.), p. 395. Academic Press, New York.
Hjelm, K. K. (1970). *Exp. Cell Res.* **60,** 191.
Holz, G. G., Jr. (1960). *Biol. Bull.* **118,** 84.
Holz, G. G., Jr. (1964). *In* "Biochemistry and Physiology of Protozoa" (S. H. Hunter, ed.), Vol. 3, p. 199. Academic Press, New York.

Holz, G. G., Jr., Scherbaum, O. H., and Williams, N. (1957). *Exp. Cell Res.* **13,** 618.
Holz, G. G., Jr., Erwin, J. A., and Davis, R. J. (1959). *J. Protozool.* **6,** 149.
Honigberg, B. M., Balamuth, W., Bovee, E. C., Corliss, J. O., Gojdics, M., Hall, R. P., Kudo, R. R., Levine, N. D., Loeblich, A. R., Jr., Weiser, J., and Wenrich, D. H. (1964). *J. Protozool.* **11,** 7.
Huddleston, M. S., Cavalier, L., Elliasen, J., Kozak, M., and Maledon, A. (1964). *Life Sci.* **3,** 1181.
Hufnagel, L. A. (1968). *J. Cell Biol.* **39,** 63A.
Hwang, S. W., Davis, E. E., and Alexander, M. T. (1964). *Science* **144,** 64.
Ito, J., Lee, Y. C., and Scherbaum, O. H. (1968). *Exp. Cell Res.* **53,** 85.
Jahn, T. L., and Brown, M. (1961). *Amer. Zool.* **1,** 454.
Jahn, T. L., Brown, M., and Winet, H. (1961). *Amer. Zool.* **1,** 454.
James, T. W. (1961). *Pathol. Biol.* **9,** 510.
James, T. W., and Read, C. P. (1957). *Exp. Cell Res.* **13,** 510.
Jones, R. F., and Baker, H. G. (1946). *Nature* (*London*) **157,** 554.
Kidder, G. W. (1967). *In* "Chemical Zoology" (M. Florkin and B. T. Scheer, eds.), Vol. 1 (G. W. Kidder, ed.), p. 93. Academic Press, New York.
Kidder, G. W., and Dewey, V. C. (1951). *In* "Biochemistry and Physiology of Protozoa" (A. Lwoff, ed.), Vol. 1, p. 324. Academic Press, New York.
Kidder, G. W., and Dewey, V. C. (1957). *Arch. Biochem. Biophys.* **66,** 486.
Kimball, R. F. (1964). *In* "Biochemistry and Physiology of Protozoa" (S. H. Hutner, ed.), Vol. 3, p. 243. Academic Press, New York.
Kozloff, E. N. (1956). *J. Protozool.* **3,** 204.
Kozloff, E. N. (1962). *J. Protozol.* **9,** Suppl. 17.
Lee, K-H. (1959). *J. Amer. Pharm. Ass. Pract. Pharm. Ed.* **48,** 468.
Lee, Y. C., and Byfield, J. E. (1969). *J. Cell Biol.* **43,** 78A.
Lee, Y. C., and Scherbaum, O. H. (1965). *Nature* (*London*) **208,** 1350.
Levy, M. R., Gollon, C. E., and Elliott, A. M. (1969). *Exp. Cell Res.* **55,** 295.
Loefer, J. B., and Mefferd, R. B., Jr. (1952). *Amer. Natur.* **86,** 325.
Lwoff, A. (1923). *C. R. Acad. Sci. Ser.* **D176,** 928.
Mackenzie, T. B., Stone, G. E., and Prescott, D. M. (1966). *J. Cell Biol.* **31,** 633.
Mandel, M. (1967). *In* "Chemical Zoology" (M. Florkin and B. T. Scheer, eds.), Vol. 1 (G. W. Kidder, ed.), p. 541. Academic Press, New York.
Maupas, E. (1883). *Arch. Zool. Exp. Gen. Ser. 2* **1,** 427.
Maupas, E. (1888). *Arch. Zool. Exp. Gen. Ser. 2* **6,** 165.
Metz, C. B., and Westfall, J. A. (1954). *Biol. Bull.* **107,** 106.
Miller, O. L., Jr., and Stone, G. E. (1963). *J. Protozool.* **10,** 280.
Mita, T., Shiomi, H., and Iwai, K. (1966). *Exp. Cell Res.* **43,** 696.
Müller, M. (1967). *In* "Chemical Zoology" (M. Florkin and B. T. Scheer, eds.), Vol. 1 (G. W. Kidder, ed.), p. 351. Academic Press, New York.
Müller, M. (1970). *J. Protozool.* **17,** Suppl. 13.
Müller, M., and Rohlich, P. (1961). *Acta Morphol. Acad. Sci. Hung.* **10,** 297.
Müller, M., Baudhuin, P., and de Duve, C. (1966). *J. Cell Physiol.* **68,** 165.
Müller, M., Törö, I., Polgar, M., and Druga, A. (1963). *Acta Biol.* (*Budapest*) **14,** 209.
Muramatsu, M. (1970). *In* "Methods in Cell Physiology" (D. M. Prescott, ed.), Vol. 4, p. 195. Academic Press, New York.
Nachtwey, D. S. (1961). *Diss. Abstr.* **B22,** 1758.
Nanney, D. L. (1966a). *Amer. Natur.* **100,** 303.
Nanney, D. L. (1966b). *Genetics* **54,** 955.

Nanney, D. L. (1966c). *J. Protozool.* **13,** 483.
Nanney, D. L. (1966d). *J. Exp. Zool.* **161,** 307.
Nanney, D. L. (1967a). *J. Protozool.* **14,** 690.
Nanney, D. L. (1967b). *J. Exp. Zool.* **166,** 163.
Nanney, D. L. (1968a). *Science* **160,** 496.
Nanney, D. L. (1968b). *J. Protozool.* **15,** 109.
Nanney, D. L. (1970). *J. Exp. Zool.* **175,** 383.
Nanney, D. L. (1971). *J. Protozool.* **18,** 33.
Nilsson, J. R. (1968). *J. Protozool.* **15,** Suppl. 29.
Nilsson, J. R. (1969). *J. Protozool.* **16,** Suppl. 32.
Nilsson, J. R., and Chapman-Andresen, C. (1968). *J. Protozool.* **15,** Suppl. 37.
Nilsson, J. R., and Williams, N. E. (1966). *C. R. Trav. Lab. Carlsberg* **35,** 119.
Organ, A. E., Bovee, E. C., and Jahn, T. L. (1968). *J. Protozool.* **15,** Suppl. 23.
Organ, A. E., Bovee, E. C., and Jahn, T. L. (1969). *J. Protozool.* **16,** Suppl. 9.
Padilla, G. M., and Cameron, I. (1964). *J. Cell. Comp. Physiol.* **64,** 303.
Phelps, A. (1935). *J. Exp. Zool.* **70,** 109.
Phelps, A. (1946). *J. Exp. Zool.* **102,** 277.
Phelps, A. (1961). *Amer. Zool.* **1,** 467.
Pitelka, D. R. (1961). *J. Protozool.* **8,** 75.
Pitelka, D. R. (1969). *In* "Research in Protozoology" (T-T. Chen, ed.), Vol. 3, p. 279. Pergamon Press, Oxford.
Preer, J. R., Jr. (1969). *In* "Research in Protozoology" (T-T. Chen, ed.), Vol. 3, p. 129. Pergamon Press, Oxford.
Prescott, D. M. (1957a). *Exp. Cell Res.* **12,** 126.
Prescott, D. M. (1957b). *J. Protozool.* **4,** 252.
Prescott, D. M. (1958). *Physiol. Zool.* **31,** 111.
Prescott, D. M., and Stone, G. E. (1967). *In* "Research in Protozoology" (T-T. Chen, ed.), Vol. 2, p. 117. Pergamon Press, Oxford.
Prescott, D. M., Rao, M. V., Evenson, D. P., Stone, G. E., and Thrasher, J. D. (1966). *In* "Methods in Cell Physiology" (D. M. Prescott, ed.), Vol. 2, p. 131. Academic Press, New York.
Rasmussen, L. (1963). *C. R. Trav. Lab. Carlsberg* **33,** 53.
Rasmussen, L., and Kludt, T. A. (1970). *Exp. Cell Res.* **59,** 457.
Ray, C., Jr. (1956a). *J. Protozool.* **3,** 88.
Ray, C., Jr. (1956b). *Stain Technol.* **31,** 271.
Ron, A., and Guttman, R. (1961). *Exp. Cell Res.* **25,** 176.
Rooney, D. W., and Eiler, J. J. (1967). *Exp. Cell Res.* **48,** 649.
Rooney, D. W., and Eiler, J. J. (1969). *Exp. Cell Res.* **54,** 49.
Roque, M., de Puytorac, P., and Savoie, A. (1970). *J. Protozool.* **17,** Suppl. 37.
Roth, L. E., and Minick, O. T. (1958). *J. Protozool.* **5,** Suppl. 22.
Roth, L. E., and Minick, O. T. (1961). *J. Protozool.* **8,** 12.
Ryley, J. F. (1967). *In* "Chemical Zoology" (M. Florkin and B. T. Scheer, eds.), Vol. 1 (G. W. Kidder, ed.), p. 55. Academic Press, New York.
Sandon, H. (1927). "The Composition and Disruption of the Protozoan Fauna of the Soil." Oliver and Boyd, Edinburgh.
Satir, B., and Dirksen, E. R. (1971). *J. Cell Biol.* **48,** 143.
Sato, H., and Saito, M. (1959). *Zool. Mag.* (*Jap.*) **68,** 209.
Scherbaum, O. H. (1956). *Exp. Cell Res.* **11,** 464.
Scherbaum, O. H. (1960). *Annu. Rev. Microbiol.* **14,** 283.

Scherbaum, O. H., and Jahn, T. L. (1964). *Exp. Cell Res.* **33,** 99.
Scherbaum, O. H., and Loefer, J. B. (1964). *In* "Biochemistry and Physiology of Protozoa" (S. H. Hutner, ed.), Vol. 3, p. 10. Academic Press, New York.
Scherbaum, O. H., and Zeuthen, E. (1954). *Exp. Cell Res.* **6,** 221.
Scherbaum, O. H., and Zeuthen, E. (1955). *Exp. Cell Res.* Suppl. **3,** 312.
Schewiakoff, W. (1889). *Bibliotheca Zool.* **5,** 1.
Schmid, P. (1967a). *Exp. Cell Res.* **45,** 460.
Schmid, P. (1967b). *Exp. Cell Res.* **45,** 471.
Seaman, G. R. (1955). *In* "Biochemistry and Physiology of Protozoa" (S. H. Hutner and A. Lwoff, eds.), Vol. 2, p. 91. Academic Press, New York.
Seaman, G. R. (1961). *J. Protozool.* **8,** 204.
Seaman, G. R. (1962). *J. Protozool.* **9,** 335.
Seaman, G. R., and Mancilla, R. (1961). *Progr. Protozool. Proc. Int. Congr. Protozool.* **1,** 165.
Seaman, G. R., and Reifel, R. M. (1963). *Annu. Rev. Microbiol.* **17,** 451.
Speth, V., and Wunderlich, F. (1970). *J. Cell Biol.* **47,** 772.
Stein, F. (1867). Der Organismus der Infusionsthiere nach eigenen Forschungen in systematischer Reichenfolge bearbeitet, II, Leipzig.
Stone, G. E. (1968a). *J. Cell Biol.* **39,** 130A.
Stone, G. E. (1968b). *J. Cell Biol.* **39,** 559.
Stoner, L. C., and Dunham, P. B. (1970). *J. Protozool.* **17,** Suppl. 20.
Summers, L. G. (1963). *J. Protozool.* **10,** 288.
Swift, H., Adams, B. J., and Larsen, K. (1964). *J. Roy. Microsc. Soc.* **83,** 161.
Szyszko, A. H., Prazak, B. L., Ehret, C. F., Eisler, W. J., Jr., and Wille, J. J., Jr. (1968). *J. Protozool.* **15,** 781.
Tamura, S., Tsuruhara, T., and Watanabe, Y. (1969). *Exp. Cell Res.* **55,** 351.
Tartar, V. (1967). *In* "Research in Protozoology" (T-T. Chen, ed.), Vol. 2, p. 1. Pergamon Press, Oxford.
Thormar, H. (1962a). *Exp. Cell Res.* **27,** 585.
Thormar, H. (1962b). *Exp. Cell Res.* **28,** 269.
Tokuyasu, K., and Scherbaum, O. H. (1965). *J. Cell Biol.* **27,** 67.
van Leeuwenhoek, A. (1677). *Phil. Trans. Roy. Soc. London Ser.* **B12,** 821.
Villadsen, I. S., and Zeuthen, E. (1969). *J. Protozool.* **16,** Suppl. 31.
Wang, G-T., and Marquardt, W. C. (1966). *J. Protozool.* **13,** 123.
Whitson, G. L., Padilla, G. M., Canning, R. E., Cameron, I. L., Anderson, N. G., and Elrod, L. H. (1966). *Nat. Cancer Inst. Monogr.* **21,** 317.
Wille, J. J., Jr., and Ehret, C. F. (1968a). *J. Protozool.* **15,** 785.
Wille, J. J., Jr., and Ehret, C. F. (1968b). *J. Protozool.* **15,** 789.
Williams, N. E. (1964a). *J. Protozool.* **11,** 230.
Williams, N. E. (1964b). *J. Protozool.* **11,** 566.
Williams, N. E., and Luft, J. H. (1968). *J. Ultrastruct. Res.* **25,** 271.
Williams, N. E., and Scherbaum, O. H. (1959). *J. Embryol. Exp. Morphol.* **7,** 241.
Williams, N. E., and Zeuthen, E. (1964). *J. Protozool.* **11,** Suppl. 10.
Williams, N. E., and Zeuthen, E. (1966). *C. R. Trav. Lab. Carlsberg* **35,** 101.
Williams, N. E., Anderson, E., Kessel, R., and Beams, H. W. (1960). *J. Protozool.* **7,** Suppl. 27.
Windsor, D. A. (1960). *J. Protozool.* **7,** Suppl. 27.
Wingo, W. J., and Anderson, N. L. (1951). *J. Exp. Zool.* **116,** 571.

Wolfe, J. (1970). *J. Cell Sci.* **6,** 679.
Wunderlich, F. (1968). *J. Cell Biol.* **38,** 458.
Wunderlich, F. (1969a). *Exp. Cell Res.* **56,** 369.
Wunderlich, F., and Peyk, D. (1969). *Experentia* **25,** 1278.
Zeuthen, E., and Scherbaum, O. H. (1954). *Proc. Symp. Colston Res. Soc.* **7,** 141.

CHAPTER 2

Carbohydrate Metabolism

Introduction

Carbohydrates play an important part in the life of all organisms. Higher plants contain vast amounts of structural cellulose, and insects and Crustacea have chitin as the predominant material of their exoskeletons. For most organisms, carbohydrate metabolism is essentially the metabolism of glucose and its products. The major pathway for the breakdown of glucose is glycolysis, a process not requiring oxygen and resulting in the formation of lactate. Aerobically, glucose is converted to CO_2. Many animals, including man, use carbohydrates as a primary source of energy; and stores of carbohydrates, as glycogen, are found in the liver and muscles.

Except for the fact that *Tetrahymena* possesses a glyoxalate cycle for the conversion of fat to carbohydrate, its carbohydrate metabolism is closely related to that of mammals. Nevertheless, the breakdown of glucose by glycolysis is not ordinarily the principal source of energy for *Tetrahymena*.

Utilization of Carbohydrates

Although carbohydrates are commonly added to media in which *Tetrahymena* is grown, the organism does not require them for growth. In fact, under appropriate conditions of pH and amino acid content of the medium, population growth in a carbohydrate-free medium can be equivalent to that in the presence of glucose (Hutner, 1963; Cox *et al.*, 1968). In most cases, however, a source of carbohydrate has been found to be obligatory for optimum production of cells. This is true for several species: *Tetrahymena corlissi* (Holz, 1964; Kessler, 1961), *T. patula* (Kessler, 1961), *T. limacis* (Kessler, 1961), *T. setifera* (Holz *et al.*, 1962), *T. pyriformis* (Warnock and van Eys, 1963b), and *T. paravorax* (Holz *et al.*, 1961). Glucose is the best carbohydrate source for all except *T. corlissi* (Kessler, 1961) and *T. pyriformis* (Reynolds and Wragg, 1962), which prefer dextrin. Mannose, fructose, and galactose show moderate-to-good stimulation of growth (Warnock, 1962).

There has not always been agreement on what strains of *Tetrahymena* ferment a particular carbohydrate. From a number of reports, it is generally agreed that most strains ferment glucose, fructose, mannose, maltose, dextrin, starch, and glycogen (Kidder and Dewey, 1945; Loefer and McDaniel, 1950; Shaw and Williams, 1963; Elliott and Outka, 1956). One species, *T. vorax*, ferments lactose and galactose (Kidder and Dewey, 1945); but none utilize other disaccharides, other polysaccharides, pentoses, or polyhydric acids. Lack of ability to ferment a particular carbohydrate does not mean, however, that that compound is not a substrate for cellular enzymes. Cell-free extracts of *T. pyriformis* not only catalyze the hydrolysis of maltose and starch but also of cellobiose and sucrose (Archibald and Manners, 1959). The lack of ability to ferment the latter two compounds may be due to a permeability problem. Cell-free extracts also hydrolyze isomaltose, methyl α-D-glucoside, α-D-glucosyl phosphate, and nigerose (Archibald and Manners, 1959; Hutson and Manners, 1965). At high concentrations of maltose and isomaltose, the hydrolysis reaction may be reversed, leading to the biosynthesis of panose, 6^3-α-glucosylmaltotriose, and maltotetraose (Archibald and Manners, 1959); and in the presence of the enzyme, maltose reacts with D-glucosamine to form the disaccharide, 6-*O*-α-D-glucopyranosyl-2-amino-2-deoxy-D-glucopyranose (Tarentino and Maley, 1969). Disaccharides can also be enzymatically formed by reacting phenyl α-D-glucopyranoside with monosaccharides (Manners *et al.*, 1968). The following have been prepared: 6-*O*-α-D-glucopyranosyl-D-galactose, 6-*O*-α-D-glucopyranosyl-D-mannose, 6-*O*-α-D-glucopyranosyl-L-xylose, 1-*O*-α-D-glucopyranosylribitol, 1-*O*-α-D-glycopyranosylerythritol, 1-*O*-α-D-glucopyranosyl-D-mannitol, and a mixture of 1-*O*- and 6-*O*-α-D-glucopyranosyl-D-glucitol.

An extracellular α-amylase can be partially purified and distinguished from the internal amylase (Smith, 1961), which is found in the lysosomes of *T. pyriformis* (Müller *et al.*, 1966).

Calculations of the amount of glucose used have been made for this ciliate. An early report (Loefer, 1938) states that the amount is 2.34 to 3.92×10^{-7} mg per organism per hour, but a more recent investigation (Waithe, 1964) shows that the amount is 0.23 to 1.9×10^{-7} mg per cell per hour. Undoubtedly, many factors may influence this rate and account for the variation of values; but insulin, which influences glucose uptake in mammalian cells, has no effect on uptake by *T. pyriformis* (Waithe, 1964; Hill and van Eys, 1962). A decrease in the rate of glucose entry is apparent with increasing age of the culture (Waithe, 1964; Rogers, 1966), a result which may be due to metabolic adjustments in the cell as exogenous glucose is depleted in the culture medium. It is now known that the ability to utilize glucose greatly depends upon the medium used to grow the organism. Cells grown in the presence of glucose utilize it very well, as judged by the increase in respiratory rate, whereas those grown without glucose require supplements of sodium and potassium ions to show an equivalent increase in respiratory activity (Conner and Cline, 1967).

It is highly probable that sugar transport across the cellular membrane involves a facilitated diffusion process. When cells equilibrated with the nonutilizable pentose, L-arabinose, are exposed to glucose, there is a rapid efflux of arabinose—a phenomenon described as "counterflow" (Cirillo, 1962). The competitive nature of this process is best explained by the mobile carrier hypothesis (Park, 1961), which assumes that the cellular membrane contains mobile, stereospecific carriers which move in either direction in the membrane and at the same rate whether in combined or free form. The carrier combines with the sugar, and the net flux depends upon the total concentration of the carrier and the degree of saturation at the two surfaces. The affinity of the carrier for glucose in *T. pyriformis* must be very great, for Waithe (1964) noted that glucose utilization was independent of concentration over a range of 1.8 to 44 mM. The pinocytosis which occurs in this organism cannot account for sugar transport, as this mechanism offers no explanation for the counterflow of L-arabinose in the presence of glucose (Cirillo, 1962).

Glycogen

The reserve polysaccharide in *Tetrahymena* is glycogen, which may constitute 3–22% of the dry weight of the organism from noncarbohydrate media

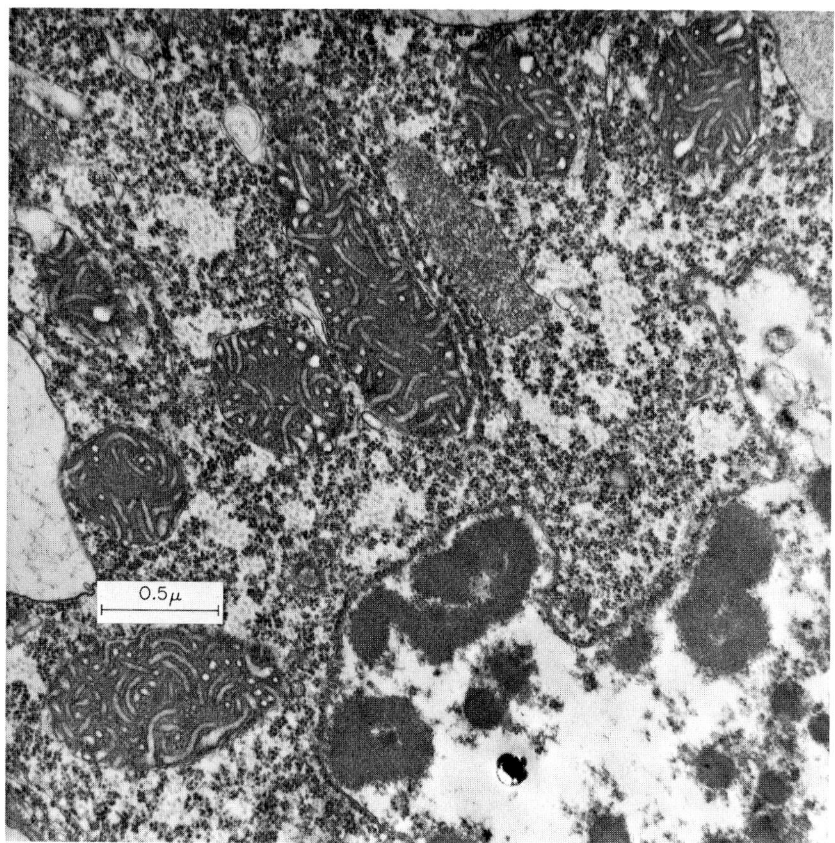

FIG. 2.1. Glycogen islands of *Tetrahymena pyriformis* HSM. Islands are less dense than surrounding polysomes. The mitochondria are typical of those in cells at early stationary phase. Magnification: ×55,000. (Courtesy of Dr. I. L. Cameron and Mr. Glenn Williams.)

(Manners and Ryley, 1952; Levy *et al.*, 1964) or up to 50% of the dry weight of young cells grown in the presence of glucose (Wagner, 1956). "Glycogen islands" can be detected by electron microscopy (Fig. 2.1). The glycogen has a unit chain length of about 13 residues, and when extracted with hot KOH has a molecular weight of 9.8×10^6 daltons (Manners and Ryley, 1952). Such harsh treatment causes disaggregation of native glycogen; but by use of the modern technique of isopycnic centrifugation in CsCl density gradients, intact glycogen may be isolated (Barber *et al.*, 1965). Prepared in this manner, the glycogen of *T. pyriformis* has a density of 1.62 to 1.65 and a particle size of 35 to 40 nm compared with a particle size of 20 nm or less for glycogen prepared by hot alkali extraction. Prepared by the isopycnic technique, rat liver glycogen exists as a continuous spectrum of particle sizes up to 200 nm, which on treatment with KOH may be broken down to 20-nm particles. Glycogens from the two sources are chemically similar in that both have α-1,4 bonds and 8–10% α-1,6 interchain linkages (Manners and Ryley, 1963). There is one report that 10–12% of the polysaccharide of *T. pyriformis* is not glycogen (Lindh and Christensson, 1962). However, no other material was noted in the CsCl density gradients.

Five types of reserve polysaccharide, all polymers of glucose, are now known for Protozoa. In addition to the glycogen of *Tetrahymena*, which is also present in *Trichomonas* (Manners and Ryley, 1955) and *Prototheca* (Manners *et al.*, 1967), there are also starch in *Polytomella* (Bourne *et al.*, 1950), *Polytoma* (Manners *et al.*, 1965), and *Chilomonas* (Hutchens *et al.*, 1948); amylopectin in rumen protozoa (Forsyth *et al.*, 1953) and *Eimeria* (Ryley and Manners, 1968); cellulose in *Acanthamoeba* (Tomlinson and Jones, 1962); and a polymer containing β-1 $\rightarrow$ 3 linkages in *Ochromonas* (Archibald *et al.*, 1958), *Astasia* (Picciolo, 1963), and *Euglena* (Kreger and Meeuse, 1952).

Low Molecular Weight Carbohydrates

There are cyclic changes in fifteen alcohol-soluble carbohydrates in synchronized cultures of *Tetrahymena pyriformis* (Whitson *et al.*, 1967). Of these carbohydrates, 2-deoxyribose, ribose, mannose, fructose, arabinose, xylose, and glucose have been identified. One unidentified component, number 13, is missing in stationary-phase cells but accumulates in cells blocked in morphogenesis. Component 15, also unidentified, is missing in log-phase cells. Washing the cells several times leaches out most of the alcohol-soluble carbohydrates; and afterward only glucose and maltose, presumably from the breakdown of glycogen, can be found. The amounts of acid-soluble hexose and pentose increase during the pretreatment for synchrony, although the total amounts of protein, ribonucleic acid, and deoxyribonucleic acid per cell parallel the cell division rate (Kamiya, 1959).

A potentially important class of carbohydrate-containing materials has been found in *T. pyriformis* (Chou and Scherbaum, 1965a,b). This consists of three acid-soluble phosphorylated amines, two of which contain deoxy sugars. All three compounds accumulate during the heat treatment used to synchronize division of the cells and may have a vital role in the process of division.

Glycolysis

Each of the enzymes involved in the interconversion of glycogen and pyruvate are present in *Tetrahymena pyriformis* (Fig. 2.2). In Table 2.1 the specific activities of the glycolytic enzymes are compared with those of ascites tumor cells, HeLa cells, chicken leukocytes, and mouse brain (Warnock and van Eys, 1962; Racker *et al.*, 1960).

Phosphofructokinase has the lowest activity, possibly due to inhibition of the enzyme in the crude supernatant used for assay. Phosphofructokinase may very well be a point of control for glycolysis, as is the case in other tissues (Lowry and Passoneau, 1964).

It is necessary to add *p*-hydroxymercuribenzoate to the reaction mixture to obtain detectable activity for fructose diphosphate phosphatase (Shrago *et al.*, 1967). Ethylenediaminetetraacetate (EDTA) is not very effective in activating the enzyme, but the reaction can be inhibited by adenosine monophosphate.

The low activity for lactate dehydrogenase in the supernatant is accounted for by the fact that this enzyme in *T. pyriformis* is bound to cellular particles (Eichel *et al.*, 1964). This is in contrast to the mammalian enzyme which is in the soluble portion of the cell. Lactate dehydrogenase

TABLE 2.1

ACTIVITY OF GLYCOLYTIC ENZYMES IN TISSUE HOMOGENATES[a]

Enzyme	*T. pyriformis*[b]	Ascites tumor[c]	HeLa cells[c]	Chicken leukocytes[c]	Mouse brain[c]
Hexokinase	50	4.8	3.7	4.7	25
Phosphohexose isomerase	20	23	—	—	—
Phosphofructokinase	0.6	5.0	7.6	14.0	51
Aldolase	12	17	6.7	15.6	16
α-Glycerophosphate dehydrogenase	3.9	0.5	—	—	—
Triose phosphate isomerase	76.3	22.7	—	—	—
3-Phosphoglyceraldehyde dehydrogenase	9.6	121	110	19	18
3-Phosphoglycerate kinase	6.0	640	700	19	111
Phosphoglycerate mutase	27	41	41	30	43
Enolase	109	27	22	21	18
Pyruvate kinase	16	138	150	55	145
Lactate dehydrogenase	0.8	230	370	111	79

[a] All values are expressed as μmoles of substrate/min/100 mg of protein.

[b] Values for *Tetrahymena* were obtained by Warnock and van Eys (1962).

[c] Values for tissues other than *Tetrahymena* are taken from Racker *et al.* (1960).

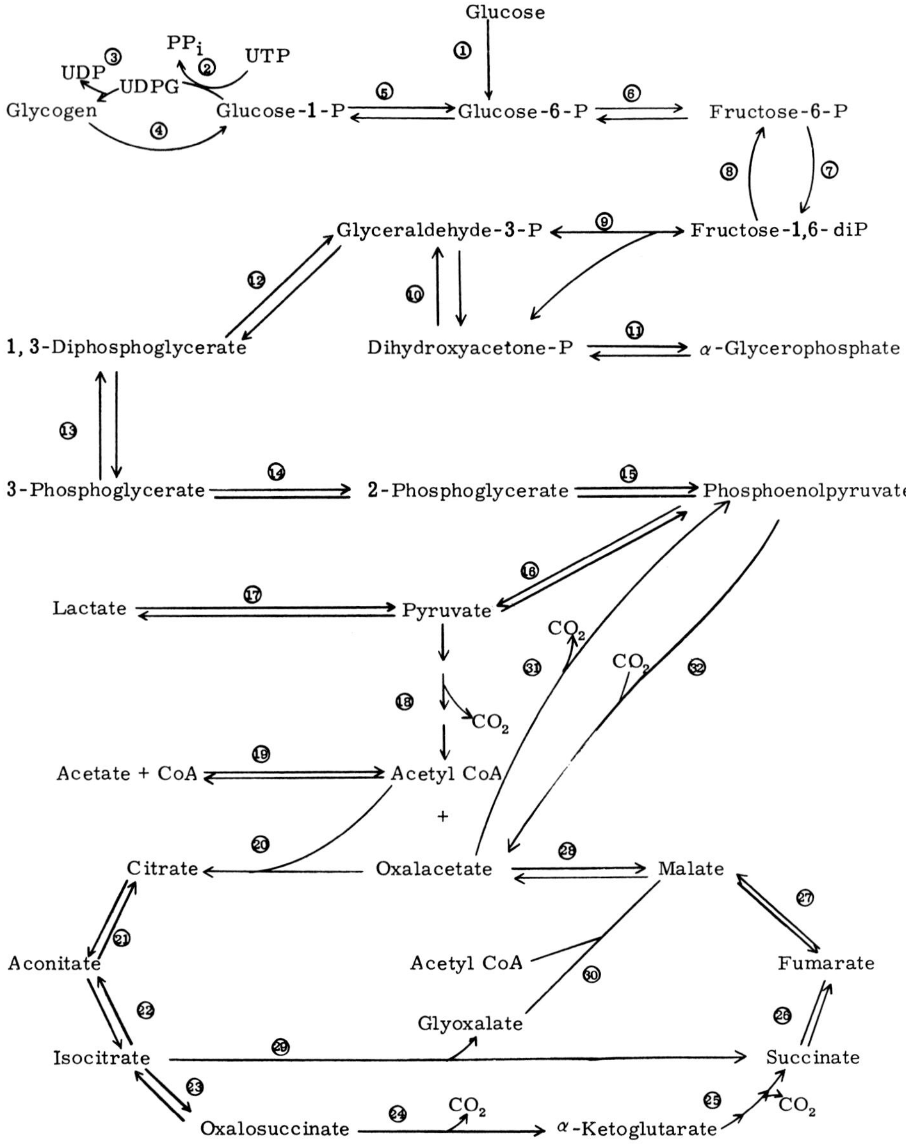

FIG. 2.2. Carbohydrate metabolism of *Tetrahymena*. Abbreviations: UDP, uridine diphosphate; UDPG, uridine diphosphate–glucose; UTP, uridine triphosphate; PP_i, inorganic pyrophosphate. Key to enzymes (circled numbers) is given on facing page.

Enzyme	Reference
1. Hexokinase	Ryley (1952); Warnock and van Eys (1962)
2. UDP–glucose pyrophosphorylase	Cook *et al.* (1968)
3. Glycogen synthase	Cook *et al.* (1968); Blum (1970)
4. Phosphorylase	Ryley (1952); Blum (1970)
5. Phosphoglucomutase	Ryley (1952); Cook *et al.* (1968)
6. Phosphohexose isomerase	Ryley (1952)
7. Phosphofructokinase	Warnock and van Eys (1962)
8. Fructose diphosphate phosphatase	Shrago *et al.* (1967)
9. Adolase	Warnock and van Eys (1962)
10. Triose phosphate isomerase	Warnock and van Eys (1962)
11. α-Glycerophosphate dehydrogenase	Warnock and van Eys (1962)
12. 3-Phosphoglyceraldehyde dehydrogenase	Warnock and van Eys (1962); Sullivan (1964)
13. 3-Phosphoglycerate kinase	Warnock and van Eys (1962)
14. Phosphoglycerate mutase	Warnock and van Eys (1962)
15. Enolase	Warnock and van Eys (1962)
16. Pyruvate kinase	Warnock and van Eys (1962); Levy and Wasmuth (1970)
17. Lactate dehydrogenase	Warnock and van Eys (1962); Eichel *et al.* (1964)
18. Pyruvate dehydrogenase	Thomas (1943); Seaman (1953a)
19. Acetyl CoA synthase	Seaman (1953b); Levy and Scherbaum (1965a)
20. Citrate synthase	Seaman (1955); Müller *et al.* (1968)
21. Aconitase	Eichel (1961); Müller *et al.* (1968)
22. Aconitase	Eichel (1961); Müller *et al.* (1968)
23. Isocitrate dehydrogenase (NAD)	Müller *et al.* (1968)
24. Isocitrate dehydrogenase (NAD)	Müller *et al.* (1968)
25. α-Ketoglutarate dehydrogenase	Seaman (1953b); Müller *et al.* (1968)
26. Succinate dehydrogenase	Ryley (1952); Roth and Eichel (1955)
27. Fumarase	Eichel (1961); Sullivan and Snyder (1962); Müller *et al.* (1968)
28. Malate dehydrogenase	Roth and Eichel (1955); Koehler and Fennell (1964)
29. Isocitrate lyase	Hogg and Kornberg (1963); Reeves *et al.* (1961)
30. Malate synthase	Hogg and Kornberg (1963)
31. Phosphoenolpyruvate carboxykinase	Shrago and Shug (1966)
32. Phosphoenolpyruvate carboxylase	Shrago *et al.* (1967)
Also present:	
Alcohol dehydrogenase	Koehler and Fennell (1964)
Glyoxylate oxidase	Müller *et al.* (1968)
Absent:	
Malate enzyme	Warnock and van Eys (1962); Shrago *et al.* (1967)
Pyruvate carboxylase	Warnock and van Eys (1962); Shrago *et al.* (1967)
Citrate cleavage enzyme	Shrago *et al.* (1967)

isozymes exist, and they vary among the different strains of *T. pyriformis* (Corbett, 1970).

Aldolase, triose phosphate isomerase, and hexokinase activities are very high; but phosphoglycerate kinase is relatively low (Warnock and van Eys, 1962). α-Glycerophosphate dehydrogenase activity is also low, as was confirmed by Helmer (1968); α-glycerophosphate is a poor substrate for the electron transport system of mitochondria isolated from *T. pyriformis* (Kobayashi, 1965).

Glycogen synthetase activity is low, but Cook *et al.* (1968) calculate that enough enzyme is present to account for the glycogen synthesis that occurs in the stationary phase. The activity is increased in cells grown in the presence of glucose and especially in the presence of glucose and theophylline (Blum, 1970). Phosphorylase activity is decreased under these conditions. Uridine diphosphate (UDP)–glucose is used as the donor for glycogen synthesis by *T. pyriformis* (Cook *et al.*, 1968). For a number of microorganisms (Greenberg and Preiss, 1964) and for corn kernels (Frydman and Cardini, 1964), adenosine diphosphate (ADP)–glucose is the donor; but for the mammalian enzyme, UDP–glucose is active (Leloir and Cardini, 1957). *Tetrahymena pyriformis* thus resembles mammals in the mechanism of glycogen synthesis.

Materials other than glycogen are degraded by the glycolytic pathway. Dextrin is cleaved, perhaps both hydrolytically and phosphorylytically, to monosaccharide units (Reynolds, 1969); and the monosaccharides, glucose, mannose, and fructose, are phosphorylated prior to their entry into the reaction sequence (Seaman, 1951a).

Glycolysis in *T. pyriformis,* as in other organisms, is blocked by iodoacetate (Conner and Cline, 1967); the sensitive step occurs at the enzyme 3-phosphoglyceraldehyde dehydrogenase. Anaerobic glycolysis can also be inhibited by oxamate (Warnock and van Eys, 1963b), a compound known to be a competitive inhibitor of lactate dehydrogenase, but only in the direction of pyruvate reduction (Novoa *et al.*, 1962). Oxamate can completely inhibit glycolysis in *T. pyriformis* either in the presence or absence of CO_2, conditions which produce one or two molecules of lactate from glucose, respectively. At a concentration of 50 m*M*, the inhibitor also blocks growth; but this effect is not reversed by pyruvate, leading to the conclusion that some other process essential to growth is sensitive to oxamate inhibition (Warnock and van Eys, 1963b).

Krebs Cycle

A number of investigators have contributed in demonstrating that all of the enzymes of the Krebs (tricarboxylic acid) cycle are present in *Tetra-*

TABLE 2.2

SPECIFIC ACTIVITIES OF MITOCHONDRIAL AND PEROXISOMAL ENZYMES IN HOMOGENATES OF *Tetrahymena pyriformis*[a]

Enzyme	Temperature (°C)	Specific activity (μmoles/min/mg)
Citrate synthase	30	67 ± 16
Aconitase	25	44 ± 13
Succinate dehydrogenase	30	63 ± 27
Fumarase	25	190 ± 67
Malate dehydrogenase	25	7900 ± 540
Isocitrate dehydrogenase (NADP)[b]	25	190 ± 50
Isocitrate lyase	30	12 ± 4.5
Malate synthase	30	20 ± 0.5
Catalase	0	30 ± 15
Lactate oxidase	25	43 ± 10
Glyoxylate oxidase	30	20

[a] Data from Müller *et al.* (1968).
[b] NADP, nicotinamide adenine dinucleotide phosphate.

hymena pyriformis (see Fig. 2.2), but the best comparative data come from Müller *et al.* (1968) (Table 2.2). In cell homogenates and also in purified mitochondria, they find all the enzymes of the cycle except isocitrate dehydrogenase [nicotinamide adenine dinucleotide (NAD)-dependent] and α-ketoglutarate dehydrogenase. Although assays for these enzymes are negative, they conclude that the enzymes must be present in the organism, for addition of either substrate to the homogenate greatly stimulates the uptake of oxygen. Multiple forms of NAD- and nicotamide adenine dinucleotide phosphate (NADP)-dependent isocitrate dehydrogenase (and NAD- and NADP-dependent malate dehydrogenase) can be seen on staining of gels after gel electrophoresis of homogenates (Allen, 1968). The absence of α-ketoglutarate dehydrogenase in homogenates conflicts with a report by Seaman (1953b) that such activity can be detected. Although the conditions of Seaman's assay were used by Müller *et al.* (1968), they were unable to reproduce his results.

The activity of several Krebs cycle enzymes has been measured for *T. pyriformis* after axenic recovery from the hemocele of female cockroaches (Seaman and Clement, 1970).

In cells that have been subjected to heat treatment in order to produce synchronous division, succinate dehydrogenase activity decreases and remains low through the first synchronous division (Sullivan and Sparks,

1961); but if comparable samples of cells are exposed to either UV or X radiation prior to assay, this decrease is not apparent (Sullivan and Sparks, 1961; Sullivan and Boyle, 1961). Similar results are reported for fumarase (Sullivan and Snyder, 1962). The succinate dehydrogenase of *T. pyriformis* is more active in young cultures than in old ones (Koehler and Fennell, 1964); and it is sensitive to malonate, β-phosphonopropionate, *trans*-1,2-cyclopentanedicarboxylate, and arsenoacetate (Seaman, 1952a,b).

An early report showed that a number of Krebs cycle intermediates were not stimulatory to growth (Kidder and Dewey, 1951), but it is possible that some of these intermediates are not transported across the cell membrane (Seaman and Houlihan, 1950).

End Products of Carbohydrate Metabolism

An early and informative study of the carbohydrate metabolism of *Tetrahymena* (Van Neil *et al.*, 1942) showed that glucose was oxidized to CO_2 in the presence of oxygen; but under anaerobic conditions succinate, lactate, and acetate were produced. Also, isotopic CO_2 was incorporated into the carboxyl groups of succinate. A later study with radioactive CO_2 (Lynch and Calvin, 1952) confirmed these results.

Under anaerobic conditions, the organism maintains its normal degree of motility at the expense of its glycogen reserves if an external supply of glucose is not present (Ryley, 1952). The glycolytic balance under these conditions involves two molecules of glucose and a molecule of CO_2 being converted to two molecules of succinate and one each of lactate and acetate (Warnock and van Eys, 1962). In the absence of CO_2, normal glycolysis is followed, with two molecules of lactate being produced for each molecule of glucose utilized (Warnock and van Eys, 1962). This information is summarized in the following reactions.

1. Aerobic	Glucose + 6 $O_2 \rightarrow$ 6 CO_2 + 6 H_2O
2. Anaerobic	2 Glucose + $CO_2 \rightarrow$ 2 succinate + lactate + acetate
3. Anaerobic without CO_2	Glucose $\rightarrow$ 2 lactate

The mechanism for incorporation of CO_2 could involve phosphoenolpyruvate carboxykinase (Warnock and van Eys, 1962), but it is more likely that the recently discovered phosphoenolpyruvate carboxylase is involved (Shrago *et al.*, 1967). In either case the product is oxalacetate. An external source of CO_2 is not required for growth (Welch, 1959).

Warnock (1962) proposed a sequence of reactions involved in the production of succinate, lactate, and acetate from glycogen. These are summarized as follows.

$$2\ \text{Glycogen} + 4\ P_i + 4\ NAD^+ + 2\ ADP \rightarrow 4\ \text{phosphoenolpyruvate} + 2\ ATP + 4\ NADH\ 6\ H^+ \quad (2.1)$$
$$2\ \text{Phosphoenolpyruvate} + 2\ ADP + 2\ H^+ \rightarrow 2\ \text{pyruvate} + 2\ ATP \quad (2.2)$$
$$1\ \text{Pyruvate} + NADH + H^+ \rightarrow 1\ \text{lactate} + NAD^+ \quad (2.3)$$
$$1\ \text{Pyruvate} + NAD^+ \rightarrow 1\ \text{acetate} + NADH + H^+ + CO_2 \quad (2.4)$$
$$2\ \text{Phosphoenolpyruvate} + 2\ CO_2 \rightarrow 2\ \text{oxalacetate} + 2\ P_i \quad (2.5)$$
$$2\ \text{Oxalacetate} + 2\ NADH + 2\ H^+ \rightarrow 2\ \text{malate} + 2\ NAD^+ \quad (2.6)$$
$$2\ \text{Malate} \rightarrow 2\ \text{fumarate} \quad (2.7)$$
$$2\ \text{Fumarate} + 2\ NADH + 2\ H^+ \rightarrow 2\ \text{succinate} + 2\ NAD^+ \quad (2.8)$$

$$\text{Sum: } 2\ \text{Glycogen} + 4\ P_i + 4\ ADP + CO_2 \rightarrow 2\ \text{succinate} + 1\ \text{lactate} + 1\ \text{acetate} + 4\ ATP \quad (2.9)$$

The fact that malonic acid, known to be an inhibitor of succinate dehydrogenase, leads to increased production of lactic acid in either aerobic or anaerobic systems was given as evidence that succinate production involves reversal of three reactions of the Krebs cycle (Warnock and van Eys, 1962). This reversal process would provide a means of reoxidizing the reduced NAD formed during glycolysis; the conversion of two molecules of oxalacetate to succinate utilizes four such molecules.

Although this sequence of reactions is attractive, no NADH-dependent fumarate reductase [reaction (2.8)] is found in *T. pyriformis* (Rahat *et al.*, 1964). From an investigation of the NADH oxidase activity of this organism and from the discovery of a malate–oxalacetate shuttle for the oxidation of extramitochondrial NADH, the following sequence is proposed to replace reactions (2.6), (2.7), and (2.8) (Rahat *et al.*, 1964).

$$4\ \text{Oxalacetate} + 4\ NADH_0 + 4\ H^+ \rightarrow 4\ \text{malate} + 4\ NAD_0^+ \quad (2.10)$$
$$2\ \text{Malate} \rightarrow 2\ \text{fumarate} + 2\ H_2O \quad (2.11)$$
$$2\ \text{Malate} + 2\ NAD_1^+ \rightarrow 2\ \text{oxalacetate} + 2\ NADH_1 \quad (2.12)$$
$$2\ NADH_1 + 2\ Q^+ \rightarrow 2\ NAD_1 + 2\ QH \quad (2.13)$$
$$2\ \text{Fumarate} + 2\ QH \rightarrow 2\ \text{succinate} + 2\ Q^+ \quad (2.14)$$

$$\text{Sum: } 2\ \text{Oxalacetate} + 4\ NADH_0 + 4\ H^+ \rightarrow 2\ \text{succinate} + 4\ NAD_0^+ + 2\ H_2O \quad (2.15)$$

In this sequence, NAD_0^+ refers to cytoplasmic NAD^+, NAD_1^+ signifies mitochondrial NAD^+, and Q represents an undefined acceptor. Substitution of reactions (2.11) to (2.14) for reactions (2.6) to (2.8) leads to the same summarized scheme [reaction (2.9)]. The effect of malonate on this sequence is explained by the fact that, at the concentration used, it inhibited malate dehydrogenase.

In the absence of oxygen and carbon dioxide, the above sequence stops with the production of four molecules of lactate and with the regeneration of four molecules of NAD^+ by lactate dehydrogenase.

Phosphogluconate Oxidative Pathway

It is quite certain that *Tetrahymena* have at least part of the phosphogluconate oxidative pathway, also known as the hexose monophosphate shunt. This pathway provides reduced NADP for fatty acid biosynthesis and furnishes ribose for nucleic acid production. One report regarding the enzymes of this series states that glucose 6-phosphate dehydrogenase is present and that the product, 6-phosphogluconate, is further converted to ribulose 5-phosphate, as would be expected if the pathway is operative (Seaman, 1951b). The same report states that glucose oxidase (dehydrogenase) activity can also be detected; this reaction produces gluconate, which can be phosphorylated to 6-phosphogluconate. Multiple forms of the glucose 6-phosphate dehydrogenase of *T. pyriformis* can be detected on staining of gels after gel electrophoresis of cell extracts (Allen, 1968). Transaldolase and transketolase may not be present; Holz (1964) quotes Hogg as saying that the labeling pattern of glycogen formed from ^{14}C-acetate does not show that these enzymes are very active.

Crabtree and Pasteur Effects

The inhibition of respiration by glucose is called the Crabtree effect. *Tetrahymena pyriformis* demonstrates such an effect, but it is pH-dependent and also dependent on the presence of CO_2 (Warnock and van Eys, 1962). This implies that, under the appropriate conditions, the cell will utilize more glucose metabolized via the glycolytic pathway for the production of ATP, and will use the respiratory system less.

The opposite of the Crabtree effect, the inhibition of glycolysis by oxygen, is called the Pasteur effect. This inhibition is strong in *T. pyriformis* (Warnock and van Eys, 1962; 1963a). Respiration is not normally at the expense of the intracellular glycogen reserves; glycogen actually increases aerobically in the absence of nutrients, indicating a reversal of glycolysis. Also, fewer glycolytic products are formed aerobically than anaerobically.

Conversion of Lipid to Carbohydrate

The respiratory quotient of *Tetrahymena pyriformis* is low, implying that lipid, rather than carbohydrate, is oxidized to supply cellular adenosine triphosphate (ATP) (Ryley, 1952; Wagner, 1956). Whereas glucose or any of a number of dicarboxylic acids give only a slight stimulation of respiration, acetate, butyrate, and propionate almost double the rate (Ryley, 1952). Lipid thus appears more important than carbohydrate for energy production.

For this organism, an interesting relationship exists between lipid and carbohydrate metabolism. As stated before, glycogen can be synthesized under aerobic conditions. This occurs even in a nonnutritive medium, and glycogen accumulates at the expense of intracellular lipids (Wagner, 1956; Warnock and van Eys, 1962). Addition of either glucose or lipids leads to an increase in glycogen production; but with the exception of phenylalanine, a ketogenic compound, amino acids are not effective in this regard (Hogg and Wagner, 1956).

When first discovered, the fact that glycogen could be produced from lipid was startling because there was no known mechanism for accomplishing this. Obviously, there can be no net synthesis of glycogen from fat (acetate), which is metabolized by the Krebs cycle, since two molecules of CO_2 are produced for each molecule of acetate utilized. This is the situation in mammalian cells. But it was apparent to workers in this field that some mechanism for the conversion of fat to glycogen must exist. A sizable portion of ^{14}C-acetate was converted to glucose, and carbons 3 and 4 of glucose contained 96% of the activity (Hogg, 1963). After only 10 min, stationary-phase cells converted ^{14}C-acetate primarily to malate and aspartate, compounds which can lead to glycogen if a mechanism for carboxylating phosphoenolpyruvate is available. In contrast, citrate and glutamate were most highly labeled by log-phase cells, an indication that the normal Krebs cycle is operative at this stage of growth (Levy and Scherbaum, 1965a).

The first mechanism introduced to account for the production of glycogen from acetate was the formation of succinate from two molecules of acetyl CoA (Seaman and Naschke, 1955). The supposition that succinate is synthesized by such a reaction is now considered to be invalid, since the glyoxalate cycle enzymes, isocitrate lyase and malate synthase, are known to be present in this organism (Hogg, 1959; Reeves *et al.*, 1961) (see Fig. 2.2). These enzymes produce succinate and malate, which may be converted to oxalacetate by succinate dehydrogenase, fumarase, and malate dehydrogenase and subsequently to phosphoenolpyruvate by phosphoenolpyruvate carboxykinase. By reversal of glycolysis, phosphoenolpyruvate is converted to glycogen. The production of phosphoenolpyruvate from acetyl CoA is summarized as follows.

Acetyl CoA + oxalacetate	→ citrate + CoA	(2.16)
Citrate	→ isocitrate	(2.17)
Isocitrate	→ glyoxalate + succinate	(2.18)
Glyoxalate + acetyl CoA	→ malate + CoA	(2.19)
Succinate + FAD	→ fumarate + $FADH_2$	(2.20)
Fumarate	→ malate	(2.21)
2 Malate + 2 NAD^+	→ 2 oxalacetate + 2 NADH + 2 H^+	(2.22)
Oxalacetate + ATP	→ phosphoenolpyruvate + CO_2 + ADP	(2.23)

Sum: 2 Acetyl CoA + FAD + ATP + 2 NAD^+ → phosphoenolpyruvate + CO_2 + ADP + $FADH_2$ + 2 NADH + 2 H^+ (2.24)

The process is cyclic if succinate is considered as an end product, so that one molecule of malate is converted to a molecule of oxalacetate. Succinate is indeed an end product of anerobic glycogenolysis in *T. pyriformis* [reaction (2.15)]; but it is unlikely that this succinate is formed by isocitrate lyase, since such a scheme would result in the formation of much reduced NAD. It is difficult to see how reduced NAD could be utilized under anaerobic conditions.

Considerable evidence has now accumulated to show that the glyoxalate cycle enzymes are intimately involved in protozoan glyconeogenesis. These enzymes are not found in mammalian tissues, but they are present in germinating fatty seedlings (Marcus and Velasco, 1960), algae (Callely and Lloyd, 1964), *Acanthamoeba* (Tomlinson, 1967), *Neurospora* (Kobr *et al.*, 1969), and yeasts (Barnett and Kornberg, 1960). When *T. pyriformis* is grown in proteose–peptone medium, the cells contain malate synthase in high activity but only small amounts of isocitrate lyase (Hogg and Kornberg, 1963; Reeves *et al.*, 1961). When glucose is added to this medium, the formation of malate synthase is repressed; this repression is partially relieved by acetate. In most strains of *T. pyriformis,* the formation of isocitrate lyase is stimulated by acetate in the presence or absence of glucose (Hogg and Kornberg, 1963; Parsons and Kemper, 1968). Cells grown in chemically defined media containing acetate have both enzymes in high activity; if glucose is present in the medium, the activities are extremely low. Hogg and Kornberg interpret these data to mean that although the glyoxalate cycle enzymes are operationally linked, they are not coordinately repressible.

Both isocitrate lyase and malate synthase are present when needed to produce glycogen from acetate. When gluconeogenesis is increased during the stationary phase or during the static conditions induced by low oxygen tension, increase in the total amount of enzyme may not be a critical factor, for inhibitors of ribonucleic acid and protein synthesis suppress the increase in enzyme activity but the glyconeogenic rate is undiminished. The increased capacity for glycogen synthesis in stationary cells would thus appear to depend primarily on an activation, rather than an increase, of the glyoxalate bypass enzymes (Levy *et al.*, 1964; Levy, 1967a,b). The activities of acetyl CoA synthase and malate dehydrogenase do not change when the stationary phase is introduced; apparently the only enzymes affected are isocitrate lyase and malate synthase (Levy and Scherbaum, 1965a). However, the intracellular concentration of acetyl CoA drops (Chua and Ronkin, 1967).

In *T. pyriformis*, the glyoxylate cycle enzymes are found primarily in one of the particulate fractions. This fraction also contains catalase, and the particles are called "peroxisomes" (Müller and Hogg, 1967). In castor bean seedlings, the glyoxylate cycle enzymes are also associated with a particle containing catalase (Breidenbach and Beevers, 1967).

The peroxisomes of *T. pyriformis* have a median equilibrium density in aqueous sucrose of 1.24 to 1.25 and a median sedimentation coefficient in 0.25 *M* sucrose of 4 to 5 $\times$ 10^3 S. They contain most of the catalase, D-amino acid oxidase, L-α-hydroxy acid oxidase, glyoxylate oxidase, isocitrate lyase, and malate synthase activities and about half of the NADP-linked isocitrate dehydrogenase, but they comprise only about 10% of the total cellular protein (Müller *et al.*, 1968; Müller, 1969). Acetyl CoA synthase is also associated with the peroxisome, at least under conditions that favor an active glyoxalate bypass (Levy, 1970). The activity of L-α-hydroxy acid oxidase in the peroxisome does not correlate with the capacity for gluconeogenesis but is high under conditions of near anaerobiosis (Levy and Hunt, 1967; Levy and Wasmuth, 1970). It appears that the production of glycogen from fat requires high activities of the glyoxalate cycle enzymes, which are incorporated into an organized intracellular structure, the peroxisome, and which are selectively affected under the physiological conditions necessary for glyconeogenesis.

An explanation has been proposed for the physiological significance of glyconeogenesis under conditions of low oxygen tension (Levy and Scherbaum, 1965b; Hogg, 1969). It is that, in the natural habitat of *Tetrahymena*, photosynthetic organisms cease oxygen production during the night. The falling oxygen tension may lead to an increase in glyconeogenesis. This makes possible survival of the organism under temporary anaerobic conditions, where metabolism depends on glycogen fermentation.

Other Aspects of Glyconeogenesis

Data have been presented to show that phosphoenolpyruvate carboxykinase is involved in glyconeogenesis (Shrago and Shug, 1966; Shrago *et al.*, 1967). The activity of this enzyme is very high in cells grown on proteose–peptone, but it is lower for cells grown on proteose–peptone plus glucose. Enzyme activity is distributed between the nuclei, the mitochondria, and the 100,000 *g* supernatant. However, only the activity in the supernatant is markedly decreased in the presence of glucose. The enzyme is not stimulated by acetyl CoA, as is the case for the enzyme of *Escherichia coli* (Canovas and Kornberg, 1965).

The high activity of phosphoenolpyruvate carboxykinase, coupled with the high activity of malate dehydrogenase, in the soluble portion of the cell gives a pathway for phosphoenolpyruvate synthesis from dicarboxylic acids (Shrago *et al.*, 1967) which is similar to that in mammalian liver (Shrago and Lardy, 1966). Antimycin A, which inhibits electron transport in mammals but which has little effect on the electron transport of *Tetrahymena pyriformis*, stimulates the growth of the ciliate; and one factor in the increased growth is the increase in phosphoenolpyruvate carboxykinase in the soluble portion of the cell (Shug *et al.*, 1968). In addition to the repression by glucose, the availability of oxygen to the cell also regulates the enzyme activity (Shrago *et al.*, 1967). Well aerated cells of *T. pyriformis* have much less soluble phosphoenolpyruvate carboxykinase activity, an effect which is not evident in cultures of *E. coli* (Shrago and Shug, 1969). The activity of the mitochondrial enzyme is not influenced by glucose or by oxygen. This enzyme could possibly function in the synthesis of phosphoenolpyruvate for benzoquinone intermediates in the pathway for production of ubiquinone (Shrago *et al.*, 1967).

Tests for several other enzymes possibly involved in glyconeogenesis have been made (Shrago *et al.*, 1967). The phosphoenolpyruvate carboxylase which is present probably functions as the primary CO_2 fixation mechanism in this organism. Pyruvate carboxylase, which is an important glyconeogenic enzyme in mammals, is not present. The NADP-linked malate enzyme, which is related to lipid production in mammals (Young *et al.*, 1964), and the citrate cleavage enzyme, which could be a source of oxalacetate, are also not active. Two other enzymes also related to glyconeogenesis, phosphoglucomutase and UDP–glucose pyrophosphorylase, are elevated during glycogen production (Cook *et al.*, 1968).

Gluconeogenesis in *T. pyriformis* is inhibited by clofibrate [ethyl-2-(*p*-chlorophenoxy)-2-methylpropionate], a hypolipidemic agent for mammals (Blum and Wexler, 1968). For the ciliate, clofibrate inhibits growth and depletes cellular glycogen. After a 17-hr exposure to the drug, the isocitrate lyase activity drops 30%, and the NADP-dependent isocitrate dehydrogenase activity rises 30% (Blum and Wexler, 1968). The investigators postulate that clofibrate either inhibits gluconeogenesis or activates glycogenolysis.

Some adrenergically reactive drugs deplete *T. pyriformis* of glycogen (Blum, 1968). Isocitrate lyase activity is decreased in these cells. However, one of these drugs, aminophylline, increases the glycogen content (Blum, 1969).

REFERENCES

Allen, S. L. (1968). *Ann. N. Y. Acad. Sci.* **151,** 190.

Archibald, A. R., and Manners, D. J. (1959). *Biochem. J.* **73,** 292.

Archibald, A. R., Manners, D. J., and Ryley, J. F. (1958). *Chem. Ind.* (*London*) p. 1516.

Barber, A. A., Harris, W. W., and Padilla, G. M. (1965). *J. Cell Biol.* **27,** 281.

Barnett, J. A., and Kornberg, H. L. (1960). *J. Gen. Microbiol.* **23,** 65.

Blum, J. J. (1968). *Mol. Pharmacol.* **4,** 247.

Blum, J. J. (1969). *Fed. Proc. Fed. Amer. Soc. Exp. Biol.* **28,** 838.

Blum, J. J. (1970). *Arch. Biochem. Biophys.* **137,** 65.

Blum, J. J., and Wexler, J. P. (1968). *Mol. Pharmacol.* **4,** 155.

Bourne, E. J., Stacey, M., and Wilkinson, I. A. (1950). *J. Chem. Soc.* p. 2694.

Breidenbach, R. W., and Beevers, H. (1967). *Biochem. Biophys. Res. Comm.* **27,** 462.

Callely, A. G., and Lloyd, D. (1964). *Biochem. J.* **90,** 483.

Canovas, J. L., and Kornberg, H. L. (1965). *Biochim. Biophys. Acta* **96,** 169.

Chua, A., and Ronkin, R. R. (1967). *Comp. Biochem. Physiol.* **21,** 425.

Chou, S. C., and Scherbaum, O. H. (1965a). *Exp. Cell Res.* **39,** 346.

Chou, S. C., and Scherbaum, O. H. (1965b). *Exp. Cell Res.* **40,** 217.

Cirillo, V. P. (1962). *J. Bacteriol.* **84,** 754.

Conner, R. L., and Cline, S. G. (1967). *J. Protozool.* **14,** 22.

Cook, D. E., Rangaraj, N. I., Best, N., and Wilken, D. R. (1968). *Arch. Biochem. Biophys.* **127,** 72.

Cox, D., Frank, O., Hutner, S. H., and Baker, H. (1968). *J. Protozool.* **15,** 713.

Corbett, J. J. (1970). *J. Protozool.* **17,** 181.

Eichel, H. J. (1961). *J. Protozool.* **8,** Suppl. 16.

Eichel, H. J., Goldenberg, E. K., and Rem, L. T. (1964). *Biochim. Biophys. Acta* **81,** 172.

Elliott, A. M., and Outka, D. E. (1956). *Biol. Bull.* **111,** 301.

Forsyth, G., Hirst, E. L., and Oxford, A. E. (1953). *J. Chem. Soc.* p. 2030.

Frydman, R. D., and Cardini, C. E. (1964). *Biochem. Biophys. Res. Commun.* **14,** 353.

Greenberg, E., and Preiss, J. (1964). *J. Biol. Chem.* **239,** PC4314.

Helmer, E. (1968). *J. Protozool.* **15,** Suppl. 36.

Hill, D. L. and van Eys, J. (1962). Unpublished results.

Hogg, J. F. (1959). *Fed. Proc. Fed. Amer. Soc. Exp. Biol.* **18,** 247.

Hogg, J. F. (1963). *Biochem. J.* **89,** 88.

Hogg, J. F. (1969). *Ann. N.Y. Acad. Sci.* **168,** 281.

Hogg, J. F., and Kornberg, H. L. (1963). *Biochem. J.* **86,** 462.

Hogg, J. F., and Wagner, C. (1956). *Fed. Proc. Fed. Amer. Soc. Exp. Biol.* **15,** 275.

Holz, G. G., Jr. (1964). *In* "Biochemistry and Physiology of Protozoa" (S. H. Hutner, ed.), Vol. 3, p. 223. Academic Press, New York.

Holz, G. G., Jr., Erwin, J. A., and Wagner, B. (1961). *J. Protozool.* **8,** 297.
Holz, G. G., Jr., Erwin, J., Wagner, B., and Rosenbaum, N. (1962). *J. Protozool.* **9,** 359.
Hutchens, J. O., Podolsky, B., and Morales, M. F. (1948). *J. Cell. Comp. Physiol.* **32,** 117.
Hutner, S. H. (1963). *In* "Progress in Protozoology" (J. Ludvik *et al.*, eds.), pp. 135–136. Academic Press, New York.
Hutson, D. H., and Manners, D. J. (1965). *Biochem. J.* **94,** 783.
Kamiya, T. (1959). *J. Biochem.* (*Tokyo*) **46,** 1187.
Kessler, D. (1961). M.S. Thesis. Syracuse University.
Kidder, G. W., and Dewey, V. C. (1945). *Physiol. Zool.* **18,** 136.
Kidder, G. W., and Dewey, V. C. (1951). *In* "Biochemistry and Physiology of Protozoa" (A. Lwoff, ed.), Vol. 1, p. 335. Academic Press, New York.
Kobayashi, S. (1965). *J. Biochem.* (*Tokyo*) **58,** 444.
Kobr, M. J., Vanderhaeghe, F., and Combépine, G. (1969). *Biochem. Biophys. Res. Comm.* **37,** 640.
Koehler, L. D., and Fennell, R. A. (1964). *J. Morphol.* **114,** 209.
Kreger, D. R., and Meeuse, B. J. D. (1952). *Biochim. Biophys. Acta* **9,** 699.
Leloir, L. F., and Cardini, C. E. (1957). *J. Amer. Chem. Soc.* **79,** 6340.
Levy, M. R. (1967a). *J. Cell. Physiol.* **69,** 247.
Levy, M. R. (1967b). *Comp. Biochem. Physiol.* **21,** 291.
Levy, M. R. (1970). *Biochem. Biophys. Res. Comm.* **39,** 1.
Levy, M. R., and Hunt, R. E. (1967). *J. Cell Biol.* **34,** 981.
Levy, M. R., and Scherbaum, O. H. (1965a). *Arch. Biochem. Biophys.* **109,** 116.
Levy, M. R., and Scherbaum, O. H. (1965b). *J. Gen. Microbiol.* **38,** 221.
Levy, M. R., and Wasmuth, J. J. (1970). *Biochem. Biophys. Acta* **201,** 205.
Levy, M. R., Scherbaum, O. H., and Hogg, J. F. (1964). *Fed. Proc. Fed. Amer. Soc. Exp. Biol.* **23,** 320.
Lindh, N. O., and Christensson, E. (1962). *Arkiv Zool.* **15,** 163.
Loefer, J. B. (1938). *J. Exp. Zool.* **79,** 167.
Loefer, J. B., and McDaniel, M. R. (1950). *Proc. Amer. Soc. Protozool.* **1,** 8.
Lowry, O. H., and Passonneau, J. V. (1964). *Arch. Exp. Pathol. Pharmakol.* **248,** 185.
Lynch, V. H., and Calvin, M. (1952). *J. Bacteriol.* **63,** 525.
Manners, D. J., and Ryley, J. F. (1952). *Biochem. J.* **52,** 480.
Manners, D. J., and Ryley, J. F. (1955). *Biochem. J.* **59,** 369.
Manners, D. J., and Ryley, J. F. (1963). *J. Protozool.* **10,** Suppl. 28.
Manners, D. J., Mercer, G. A., Stark, J. R., and Ryley, J. F. (1965). *Biochem. J.* **96,** 530.
Manners, D. J., Pennie, I. R., and Ryley, J. F. (1967). *Biochem. J.* **104,** 32P.
Manners, D. J., Pennie, I. R., and Stark, J. R. (1968). *Carbohyd. Res.* **7,** 29.
Marcus, A., and Velasco, J. (1960). *J. Biol. Chem.* **235,** 563.
Müller, M. (1969). *Ann. N.Y. Acad. Sci.* **168,** 292.
Müller, M., and Hogg, J. F. (1967). *Fed. Proc. Fed. Amer. Soc. Exp. Biol.* **26,** 284.
Müller, M., Baudhuin, P., and de Duve, C. (1966). *J. Cell. Physiol.* **68,** 165.
Müller, M., Hogg, J. F., and de Duve, C. (1968). *J. Biol. Chem.* **243,** 5385.
Novoa, W. B., Winer, A. D., Glaid, A. J., and Schwert, G. W. (1962). *J. Biol. Chem.* **234,** 1143.
Park, C. R. (1961). Membrane Transport Metab., Proc. Symp. Prague, 1960, pp. 19–21. Academic Press, New York.
Parsons, J. A., and Kemper, D. L. (1968). *J. Cell Biol.* **39,** 102A.
Picciolo, G. L. (1963). *J. Protozool.* **10,** Suppl. 9.

Rahat, M., Judd, J., and van Eys, J. (1964). *J. Biol. Chem.* **239,** 3537.

Racker, E., Wu, R., and Alpers, J. B. (1960). *In* "Amino Acids, Proteins, and Cancer Biochemistry" (J. T. Edsall, ed.), p. 175. Academic Press, New York.

Reeves, H., Papa, M., Seaman, G. R., and Ajl, S. (1961). *J. Bacteriol.* **81,** 154.

Reynolds, H. (1969). *J. Protozool.* **16,** 204.

Reynolds, H., and Wragg, J. B. (1962). *J. Protozool.* **9,** 214.

Rogers, C. G. (1966). *Can. J. Biochem.* **44,** 1493.

Roth, J. S., and Eichel, H. J. (1955). *Biol. Bull.* **108,** 308.

Ryley, J. F. (1952). *Biochem. J.* **52,** 483.

Ryley, J. F., and Manners, D. J. (1968). *J. Protozool.* **15,** Suppl. 31.

Seaman, G. R. (1951a). *J. Gen. Physiol.* **34,** 775.

Seaman, G. R. (1951b). *J. Biol. Chem.* **191,** 439.

Seaman, G. R. (1952a). *Arch. Biochem. Biophys.* **39,** 241.

Seaman, G. R. (1952b). *Arch. Biochem. Biophys.* **35,** 132.

Seaman, G. R. (1953a). *J. Bacteriol.* **65,** 744.

Seaman, G. R. (1953b). *Proc. Soc. Exp. Biol. Med.* **82,** 184.

Seaman, G. R. (1955). *In* "Biochemistry and Physiology of Protozoa" (S. H. Hutner and A. Lwoff, eds.), Vol. 2, p. 91, Academic Press, New York.

Seaman, G. R., and Clement, J. J. (1970). *J. Protozool.* **17,** 287.

Seaman, G. R., and Houlihan, R. K. (1950). *Arch. Biochem.* **26,** 436.

Seaman, G. R., and Naschke, M. D. (1955). *J. Biol. Chem.* **217,** 1.

Shaw, R. F., and Williams, N. E. (1963). *J. Protozool.* **10,** 486.

Shrago, E., and Lardy, H. A. (1966). *J. Biol. Chem.* **241,** 663.

Shrago, E., and Shug, A. L. (1966). *Biochim. Biophys. Acta* **122,** 376.

Shrago, E., and Shug, A. L. (1969). *Arch. Biochem. Biophys.* **130,** 393.

Shrago, E., Brech, W., and Templeton, K. (1967). *J. Biol. Chem.* **242,** 4060.

Shug, A. L., Ferguson, S., and Shrago, E. (1968). *Biochem. Biophys. Res. Commun.* **32,** 81.

Smith, I. (1961). Ph.D. Thesis, Columbia University, New York.

Sullivan, W. D. (1964). *J. Cell Biol.* **23,** 91A.

Sullivan, W. D., and Boyle, J. V. (1961). *Broteria Ser. Cienc. Natur.* **30,** 77.

Sullivan, W. D., and Snyder, R. L. (1962). *Exp. Cell Res.* **28,** 239.

Sullivan, W. D., and Sparks, J. T. (1961). *Exp. Cell Res.* **23,** 536.

Tarentino, A. L., and Maley, F. (1969). *Arch. Biochem. Biophys.* **130,** 80.

Thomas, J. O. (1943). Stanford University Bull. **18,** 20.

Tomlinson, G. (1967). *J. Protozool.* **14,** 114.

Tomlinson, G., and Jones, E. (1962). *Biochim. Biophys. Acta* **63,** 194.

Van Niel, C. B., Thomas, J. O., Ruben, S., and Kamen, M. D. (1942). *Proc. Nat. Acad. Sci., U.S.* **28,** 157.

Wagner, C. (1956). Ph.D. Thesis, University of Michigan, Ann Arbor, Michigan.

Waithe, W. I. (1964). *Exp. Cell Res.* **35,** 100.

Warnock, L. G. (1962). Ph.D. Thesis, Vanderbilt University, Nashville, Tennessee.

Warnock, L. G., and van Eys, J. (1962). *J. Cell. Comp. Physiol.* **60,** 53.

Warnock, L. G., and van Eys, J. (1963a). *J. Cell. Comp. Physiol.* **61,** 309.

Warnock, L. G., and van Eys, J. (1963b). *J. Bacteriol.* **85,** 1179.

Welch, C. A. (1959). *Diss. Abstr.* **20,** 1110.

Whitson, G. L., Green, J. G., Francis, A. A., and Willis, D. D. (1967). *J. Cell. Physiol.* **70,** 169.

Young, J. W., Shrago, E., and Lardy, H. A. (1964). *Biochemistry* **3,** 1687.

CHAPTER 3

Lipid Metabolism

Introduction

The lipids, a diverse group of compounds, are linked together primarily by their solubility in nonpolar solvents. Glycerides, phosphatides, and sterols are the major tissue lipids. Phosphatides and sterols are components of cellular membranes; and phosphatides, because of their dipolar nature, have been implicated in transport across the membranes. Large amounts of gycerides may be stored in the adipose tissue of mammals or in the seeds of higher plants. Glycerides represent a very concentrated source of energy; they liberate more than twice as many calories per gram as either carbohydrates or proteins. Most of this energy is derived from the catabolism of the fatty acids liberated by hydrolysis and degraded by β-oxidation.

Tetrahymena contain a variety of lipids and use to advantage the concentrated energy of glycerides, for, under usual conditions, fatty acids provide their major source of energy. The precise mechanisms by which fatty acids are formed and degraded by *Tetrahymena* have not been elucidated.

Requirements

Tetrahymena pyriformis shows no requirement for fatty acids, as mammals do, or for sterols, or, under ordinary conditions, for phospholipids (Kidder and Dewey, 1949). Apparently this species can form, from small molecules, all of the complex lipid material needed for growth. However, three other *Tetrahymena* species, *T. corlissi*, *T. paravorax*, and *T. setifera*, require sterols (Holz *et al.*, 1961a,b; 1962a). Vertebrates and higher plants make sterols; but, in addition to the three species of *Tetrahymena*, insects, Crustacea (*Artemia*), annelids (*Lumbricus*), trichomonads, some *Labyrinthula* species, and some mycoplasma also require an exogenous supply (see Hutner and Holz, 1962).

Data on the sterol requirements of the *Tetrahymena* species are summarized in Table 3.1 (Holz *et al.*, 1961a,b; 1962a). Neither species responds to mevalonate or squalene, which are aliphatic precursors of sterols. Ring-methylated products of squalene cyclizations and lanosterol demethylation are inactive, except lophenol, which is moderately effective nutritionally for *T. corlissi*. Zymosterol, a polycyclic precursor of cholesterol, satisfies *T. corlissi* and *T. paravorax*, but not *T. setifera*. More immediate cholesterol precursors, desmosterol, Δ^7-cholestenol, and 7-dehydrocholesterol are as effective as cholesterol; and a number of sterols with double bonds in the 5-position and alkyl groups on C-24 are highly active. Another study (Conner, 1959) shows that steroid hormones of vertebrates are either inactive or inhibitory. All of the potent inhibitors are of the 21-carbon or pregnane series. A hydroxyl at C-11 enhances the inhibition, but a hydroxyl at C-17 lowers the potency. The most potent are deoxycorticosterone, corticosterone, and 11-dehydrocorticosterone. The inhibition by sterols is antagonized by either stigmasterol or cholesterol. 3-Dialkylaminoethoxysterols also inhibit growth (Holmlund and Bohonos, 1966), as does diethylstilbestrol (Dewey, 1967).

The sterol requirement for *T. setifera* and *T. paravorax* is spared by L-α-dipalmitolyl phosphatidyl ethanolamine plus oleate (Erwin *et al.*, 1966; Holz *et al.*, 1962a); that for *T. corlissi* is spared, in order of decreasing activity, by DL-α-glycerophosphate, L-α-dioleolyl phosphatidyl ethanolamine, oleate, and other synthetic phosphatides (Holz *et al.*, 1961a). The sterol requirement by *T. setifera* is also spared by the fatty acids of *T. pyriformis* (Erwin *et al.*, 1966).

Tetrahymena setifera has the unusual requirement of a short-chain alcohol for growth. Ethanol or methanol, but not a wide variety of other alcohols, acids, or aldehydes, fills this requirement (Holz *et al.*, 1962a).

Phospholipids have been used with success for growing four of the more fastidious *Tetrahymena* species: *T. corlissi*, *T. limacis*, *T. patula*

TABLE 3.1

Tetrahymena Sterol Requirements[a]

Sterol	Double bonds	Substitutions	*T. corlissi* TH-X[b]	*T. paravorax* RP[c]	*T. setifera* HZ-1[d]
Lanosterol	8, 24	4α, 4β, 14α—CH_3	−	−	−
Dihydrolanosterol	8	4α, 4β, 14α—CH_3	−	−	−
4′,4′-Dimethyl-Δ^8-cholestenol	8	4α, 4β—CH_3	−	−	−
Citrostadienol	7, 24(28)	4α—CH_3, 24=$CHCH_3$	−	−	−
4α-Methyl-Δ^7-cholestenol (lophenol)	7	4α—CH_3	++	−	−
4α-Methyl-Δ^8-cholestenol	8	4α—CH_3	−	−	−
Zymosterol	8, 24		+	+	−
Desmosterol	5, 24		+++	+++	+++
Δ^7-Cholestenol	7		++	+++	+++
7-Dehydrocholesterol	5, 7		+++	+++	+++
Cholesterol	5		+++	+++	+++

Ostreasterol	5, 24(28)	$24{=}CH_2$	+++	+++	+++
Fucosterol	5, 24(28)	$24{=}CHCH_3$	+++	+++	+++
Campesterol	5	$24\alpha{-}CH_3$	+++		+++
Campesteryl acetate	5	$24\alpha{-}CH_3$		+++	+++
β-Sitosterol	5	$24\beta{-}C_2H_5$	+++	+++	+++
Clionasterol	5	$24\alpha{-}C_2H_5$	+++		+++
Brassicasterol	5, 22	$24\beta{-}CH_3$	+++		+++
Brassicasteryl acetate	5, 22	$24\beta{-}CH_3$	+++	+++	
Stigmasterol	5, 22	$24\beta{-}C_2H_5$	+++	+++	+++
Poriferasterol	5, 22	$24\alpha{-}C_2H_5$	+++	+++	+++
Ergosterol	5, 7, 22	$24\beta{-}CH_3$	+++	+++	+++
Neospongesterol	22	$24\alpha{-}CH_3$	+++	+++	
Mevalonate			−	−	−
Squalene			−	−	−

[a] *Key to symbols*: −, Does not satisfy sterol requirement for growth; +, fair source of sterol for growth; ++, good source of sterol for growth; +++, excellent source of sterol for growth.

[b] Holz *et al.*, 1961a. [c] Holz *et al.*, 1961b. [d] Holz *et al.*, 1962a.

(Kessler, 1961), and *T. vorax* (Shaw and Williams, 1963). A phospholipid requirement can be induced for *T. pyriformis* by growth at supraoptimal temperatures (Rosenbaum *et al.*, 1966). In a defined medium this organism grows at a temperature as high as 37°C; but growth is achieved at 40°C if either phosphatidyl choline or phosphatidyl ethanolamine is added. At 0° to 5°C, preparations of soy or egg lecithin (phosphatidyl choline), sitosterol, or stigmasterol plus an antioxidant permit survival for 16 to 22 weeks (Reid and Cox, 1968; Reid *et al.*, 1969; Cox, 1970). The organism does not normally remain viable at such low temperatures. The function of phospholipids for the other species is not at all clear.

A recent, illuminating investigation shows that cells grown at 40°C with phospholipids are morphologically abnormal (Erwin, 1970). The relative phospholipid content is only one-third that of normal cells. The added phospholipids are not incorporated as such into cellular membranes but rather serve as a nontoxic source of unsaturated fatty acids. Purified *T. pyriformis* fatty acids obtained from normal cells grown in a lipid-free medium and added in the presence of bovine serum albumin will support optimum growth at 40°C and prevent the production of surface abnormalities. The effects of high temperatures on growth and morphology are probably the result of inhibition of fatty acid biosynthesis and, indirectly, of the biosynthesis of phospholipids for the *T. pyriformis* cellular membrane (Erwin, 1970).

Although fatty acids are not required for growth, acetate stimulates the growth of *T. pyriformis* and is often included in the medium; propionate stimulates growth at low concentrations but inhibits growth when the amount is increased. Myristate, laurate, caprate, caprylate, caproate, and a variety of fatty acids with odd numbers of carbon atoms are all inhibitory (Dewey, 1967). Long-chain unsaturated fatty acids, such as undecylenate, oleate, α-linolenate, and γ-linolenate, cause lysis of *T. pyriformis* (Lees and Korn, 1966; Chaix and Baud, 1947). In contrast, oleate stimulates the growth of *T. corlissi* and *T. paravorax* (Holz *et al.*, 1961a,b). Oleic acid-containing detergents stimulate early phases of growth, but cell numbers become equal after 4 days—the detergents are broken, and the fatty acids are stored, making the organisms opaque and fragile (Kidder *et al.*, 1954).

Composition

Lipids comprise 8–24% of the dry weight of *Tetrahymena pyriformis* (McKee *et al.*, 1947; Taketomi, 1961; Aaronson and Baker, 1961; Carter and Gaver, 1967). The amount of lipid per cell changes with the age of the population; there is an increase during the logarithmic and early stationary phases and a decrease afterward (Everhart and Ronkin, 1966; Kozak and Huddleston, 1966). This change is reflected morphologically in that the percent of cells containing granules of neutral fat increases in the early stationary phase and then decreases (Allison and Ronkin, 1967; Thompson, 1967). Lipids accumulate in droplets located chiefly at the anterior pole of the cell (Pace and Ireland, 1945).

Tetrahymena contain a variety of lipids: glycerides, cardiolipin, phosphatidyl ethanolamine, glyceryl ethers of phosphatidyl ethanolamine, phosphatidyl choline, phosphatidyl serine, phosphatides containing aminoethyl- or aminopropylphosphonic acid, and phosphatidyl inositides (Hack *et al.*, 1962; Thompson, 1967; Erwin and Bloch, 1963; Liang and Rosenberg, 1966; Smith *et al.*, 1970; Jonah and Erwin, 1971). Contrary to an early report (Taketomi, 1961), plasmalogens are present (Thompson, 1967). Sphingomyelin is reported to be both present (Taketomi, 1961; Shorb, 1963) and absent (Thompson, 1967). Two branched-chain sphingosines are found and have been described (Carter and Gaver, 1967). These unique long-chain bases contain a total of 17 and 19 carbon atoms and have the double bond between carbons 4 and 5. They are tentatively identified as 15-methyl C_{16}-sphingosine and 17-methyl C_{18}-spingosine and are found as constituents of ceramides and ceramide aminoethylphosphonates.

Tetrahymanol, a pentacyclic triterpenoid alcohol, was discovered in *T. pyriformis* (Mallory *et al.*, 1963). This compound, which is not a sterol, is one of a class found heretofore only in some higher plats; and recently tetrahymanol has been found in the plant *Oleandra wallichii* (Zander *et al.*, 1969). The structure of tetrahymanol, at first reported differently (Mallory *et al.*, 1963), is now known to be the following, as proved by an unambiguous synthesis and by X-ray crystallographic studies (Tsuda *et al.*, 1965; Gordon and Doyne, 1966):

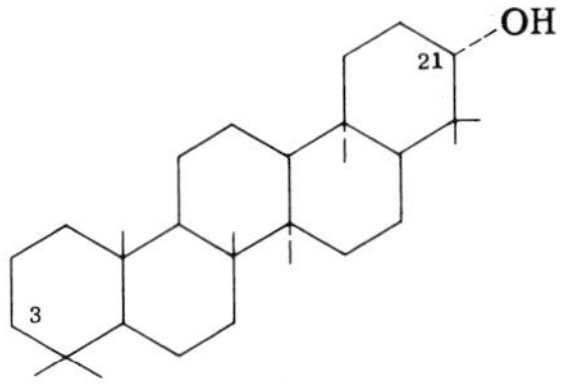

OH
21
3

A companion compound, probably another pentacyclic triterpenoid, is also present (Conner *et al.*, 1968; Jonah and Erwin, 1971). Tetrahymanol is found in the lipids of cilia and mitochondria (Jonah and Erwin, 1971).

Whether *Tetrahymena* contain sterols is in dispute. Three reports (Sea-

TABLE 3.2

FATTY ACID COMPOSITION OF *Tetrahymena* GROWN AT 25°C[a]

Fatty acid	Chain length	Double bonds	% of total			
			T. pyriformis[d]	*T. corlissi* Th-X	*T. setifera* HZ-1	*T. paravorax* RP
Laurate	C_{12}	Saturated	1.9	9.5	1.3	1.6
Isotridecanoate[b]	C_{13}	Saturated	0	0	0.8	1.9
Myristate	C_{14}	Saturated	6.5	4.9	11.4	8.7
Isopentadecanoate	C_{15}	Saturated	2.5	1.0	2.8	6.6
Pentadecanoate	C_{15}	Saturated	<1.0	0.7	0.8	0.5
Isohexadecanoate	C_{16}	Saturated	<1.0	0.5	0	0.5
Palmitate	C_{16}	Saturated	4.8	6.6	4.9	4.5

Palmitoleate	C_{16}	Δ^9	11.5	3.3	2.7	3.0
Isoheptadecanoate	C_{17}	Saturated	Trace	Trace	2.7	10.0
Isoheptadecenoate	C_{17}	?	0	0	3.4	5.0
Margarate	C_{17}	Saturated	<1.0	0.6	0	0
Heptadecenoate	C_{17}	?	<1.0	0.7	0.8	0
Stearate	C_{18}	Saturated	<1.0	1.4	0.6	1.0
Oleate	C_{18}	Δ^9	8.7	11.6	4.2	1.0
Linoleate	C_{18}	$\Delta^{9,12}$	17.9	26.1	13.2	15.5
γ-Linolenate	C_{18}	$\Delta^{6,9,12}$	37.7	30.6	47.2	33.3
Isononodecanoate	C_{19}	Saturated	0	0	2.5	4.0
Arachidate	C_{20}	Saturated	<1.0	Trace	0	Trace
Unsaturated "eicosenoate"[c]	C_{20}	?	<1.0	4.4	0.7	Trace

[a] From Erwin and Bloch (1963).
[b] The prefix "iso" refers to terminal branching of the fatty acid.
[c] A group of three unsaturated acids yielding C_{20} upon catalytic reduction.
[d] Mating type II, var. 1.

man, 1950a; Taketomi, 1961; Castrejon, 1964) state that sterols are present in the organism. One of these (Castrejon, 1964) presents evidence that cholesterol is present in young cultures and that a 3β-hydroxysterol is rapidly synthesized from ^{14}C-mevalonate. Nevertheless, no sterols could be found in *T. pyriformis* by workers who surveyed several different protozoans (Williams *et al.*, 1966; Aaronson and Baker, 1961); and no cholesterol could be found on analysis by gas chromatography (Conner and Ungar, 1964). Since the pentacyclic triterpenoids may have given a false positive test in some of the assays, it seems a fairly safe assumption at present that sterols (compounds that contain the cyclopentanophenanthrene ring) are not found in *T. pyriformis*.

Tetrahymena contain several other lipids. This organism possesses ubiquinones (Taketomi, 1961; Vakirtzi-Lemonias *et al.*, 1963; Conner *et al.*, 1968) and a material that reacts like the juvenile hormone of insects, of which the lipid farnesol may be a constituent (Schneiderman *et al.*, 1960; Schneiderman and Gilbert, 1964). There is also present an unknown polyunsaturated compound in cells grown in the light; it is not found in cells grown in the dark (Knuese and Shorb, 1966). Another unusual lipid, 1,14-docosyl disulfate, is found in *Tetrahymena*, as well as in *Ochromonas*, *Chlorella*, *Chlamydomonas*, *Pseudomonas*, and *Streptomyces griseus* (Haines, 1965). α-Tocopherol is not present (Green *et al.*, 1959).

The total extractible lipids of *T. pyriformis*, mating type II, var. 1, consist primarily of phospholipids (68%) and neutral lipids (28%). Phosphatidyl ethanolamine (54%) and phosphatidyl choline (30%) are the major phospholipids, but only small amounts of phosphatidyl serine are found (Erwin and Bloch, 1963). Glyceryl ethers of phosphatidyl ethanolamine account for a considerable amount of the total (Thompson, 1967). Triglycerides compose about 90% of the neutral lipids (Erwin and Bloch, 1963).

Drugs can modify the lipid content of the cell. Exposure of *T. pyriformis* in Ringer's solution to clofibrate [ethyl-2-(*p*-chlorophenoxy)-2-methyl propionate] causes the triglyceride content to double in 2 hr (Blum and Wexler, 1968). Exposure to promazine or chlorpromazine gives a 60% increase in the phospholipid content during the exponential phase, but there is no change in the phospholipid pattern (Rogers, 1968).

The fatty acid composition of *Tetrahymena* is qualitatively similar to that for blue-green algae and fungi (Erwin and Bloch, 1963) and yeasts (Yuan and Bloch, 1961), except for the double bond structure of the C_{18}-trienoic acid. In most lower protists and in higher plants, the octadecatrienoate is the $\Delta^{9,12,15}$ isomer (α-linolenate), whereas in *Tetrahymena* it is the $\Delta^{6,9,12}$ isomer (γ-linolenate).

The total fatty acid composition of four species of *Tetrahymena* is listed in Table 3.2 (Erwin and Bloch, 1963). More than 50% of the fatty acids are unsaturated. Myristate and palmitate are the major saturated fatty acids, whereas oleate, linoleate, and γ-linolenate are the principal unsaturated ones. The data for *T. pyriformis* generally agree with an earlier analysis (Müller *et al.*, 1959). Arachidonate is not detected in any species (Erwin and Bloch, 1963; Müller *et al.*, 1959; Aaronson *et al.*, 1963), and the report that *Tetrahymena* contain fatty acids with more than 20 carbon units (McKee *et al.*, 1947) has not been confirmed.

With increasing age of culture of mating type II, the ratio of monounsaturated fatty acids to their saturated analogs decreases; and the same effect can be produced by increasing the incubation temperature of young cultures (Erwin and Bloch, 1963).

The fatty acid composition of the various *T. pyriformis* lipids shows striking differences (Table 3.3) (Erwin and Bloch, 1963). In neutral lipids, the principal fatty acids are the saturated and monounsaturated, whereas linoleate and γ-linolenate are minor components. On the other hand, the phospholipids are rich in the two more highly unsaturated acids.

Tetrahymena paravorax and *Tetrahymena setifera* contain appreciable amounts of iso acids (Table 3.2); and isotopic evidence shows that the methyl groups of these acids are not derived from methionine, but from the isovaleryl groups in leucine metabolism (Erwin and Bloch, 1963).

The fatty acid pattern of *T. pyriformis* can be altered by feeding propionate, isobutyrate, or α-methyl-n-butyrate (Shorb, 1963). With propionate, small amounts of saturated fatty acids with 15 to 17 carbons, a monounsaturated fatty acid with 17 carbons, and a diunsaturated fatty acid with 19 carbons are present in addition to the usual ones. With a supplement of isobutyrate, iso acids containing 14–18 carbons are found; and addition of α-methyl-n-butyrate gives rise to anteiso acids with 13–19 carbons.

Metabolism of Fatty Acids and Phosphatides

Acetate, butyrate, and propionate greatly stimulate respiration of *Tetrahymena pyriformis* (Chaix *et al.*, 1947; Ryley, 1952). Apparently these short-chain fatty acids can be readily utilized, both for energy production and for incorporation into lipid and carbohydrates (Seaman, 1950b). Increased aeration decreases lipid accumulation (Engermann, 1958), and starvation conditions produce a decrease in lipids. This class of compounds is used in preference to glycogen for energy production (Wagner, 1956; Levy and Elliott, 1968).

TABLE 3.3

FATTY ACID COMPOSITION OF LIPIDS OF *Tetrahymena pyriformis* GROWN AT 37°C[a]

Lipid fraction	Relative amount of fatty acid (% of total)								Total minor fatty acids
	Myristate	Palmitate	Palmitoleate	Stearate	Oleate	Linoleate	γ-Linolenate	Arachidate	
Total lipid	20.0	15.4	7.9	2.0	7.3	15.4	19.0	2.0	11.3
Extractable lipid	18.0	15.0	6.5	3.8	8.9	16.5	20.0	2.0	5.0
Residual lipid	Trace	16.5	4.3	9.7	20.5	25.0	Trace	Trace	2.5
Free fatty acids	4.0	26.8	14.4	11.4	26.3	11.9	Trace	4.2	8.9
Neutral lipid									
Total	18.9	13.0	12.4	5.3	9.0	6.7	2.1	2.6	28.0
Triglycerides	11.5	25.0	14.4	12.9	17.1	7.9	0.8	Trace	9.3
Acetone eluate	3.0	4.8	2.4	1.1	3.3	37.8	39.0	3.0	5.3
Phospholipids									
Phosphatidyl choline	19.6	3.0	3.2	2.1	9.4	23.5	31.4	1.9	5.1
Phosphatidyl ethanolamine	14.6	12.9	7.3	1.2	12.0	22.2	19.0	Trace	6.1
Phosphatidyl serine	5.6	12.4	7.6	2.8	12.0	13.2	41.0	Trace	5.5
Total	15.3	12.3	7.2	1.0	8.5	22.8	29.5	Trace	3.2

[a] From Erwin and Bloch (1963).

β-Hydroxybutyrate dehydrogenase is present in *T. pyriformis* (Koehler and Fennell, 1964; Helmer, 1968) in the mitochondria and is nicotinamide adenine dinucleotide (NAD)-linked (Conger and Eichel, 1965). This enzyme probably serves in the breakdown of fatty acids, and there is no evidence to suggest that the process of β-oxidation differs from the β-oxidation of fatty acids in other organisms. Differential inhibition of the β-oxidation pathway by malonate and the α-oxidative sequence by imidazole suggests that both schemes are operative in *T. pyriformis* (Avins, 1968).

A number of compounds considered to be antimetabolities of acetate inhibit the growth of *T. pyriformis* (Dewey and Kidder, 1960) α-*p*-Biphenylbutyrate, α-phenylbutyrate, γ-phenylbutryate, cyclopropanecarboxylate, phenylacetate, and β, β-dimethylacrylate are all inhibitory; and the effects of all but β, β-dimethylacrylate are reversed by acetate or by short-chain fatty acids, but not by sterols or mevalonate. Other work with acetate analogs shows that α-phenylbutyrate does not delay synchronous division (Holz *et al.*, 1963) and that the effects of diphenylethylacetate can be reversed by squalene or sterols (Johnson and Jasmin, 1961). α-Biphenylbutyrate and fluoromevalonate are active in inhibiting division of synchronized *T. pyriformis* (Holz, 1963).

Very little is known about propionate metabolism in this organism. However, an enzyme that converts β-hydroxypropionate to malonic semialdehyde is active (Den *et al.*, 1959). β-Hydroxypropionate may be produced from propionate through the intermediate acrylate.

Cells grown in the presence of the unnatural fatty acids 11,14-eicosadienoate, 8,11,14-eicosatrienoate, 11-eicosenoate, or 11-octadecenoate incorporate these fatty acids into their lipids and phospholipids (Lees and Korn, 1966). Some of these can be further desaturated.

Incorporation studies using 1-^{14}C-palmitate, stearate, oleate, and linoleate show that *T. pyriformis* can elongate C_{16} to C_{18} fatty acids and can convert long-chain fatty acids to their unsaturated analogs (Erwin and Bloch, 1963). The studies led to postulation of a reaction sequence by which the organism forms trienoic acids (Fig. 3.1). The sequence

$$\text{Oleate } (C_{18}\triangle^{9}) \rightarrow \text{linoleate } (C_{18}\triangle^{9,12}) \rightarrow \gamma\text{-linolenate } (C_{18}\triangle^{6,9,12})$$

is not reversible. Compared with yeasts and green plants, the reaction is the same except for introduction of the third double bond. Eubacteria have a completely different pathway; and vertebrates have a dietary requirement for linoleate, which is converted to γ-linolenate in the same manner as for *T. pyriformis.* The ciliate pathway to polyunsaturated fatty acids thus combines, in the same organism, desaturation reactions which are characteristic of vertebrate and plant systems, respectively. Erwin and

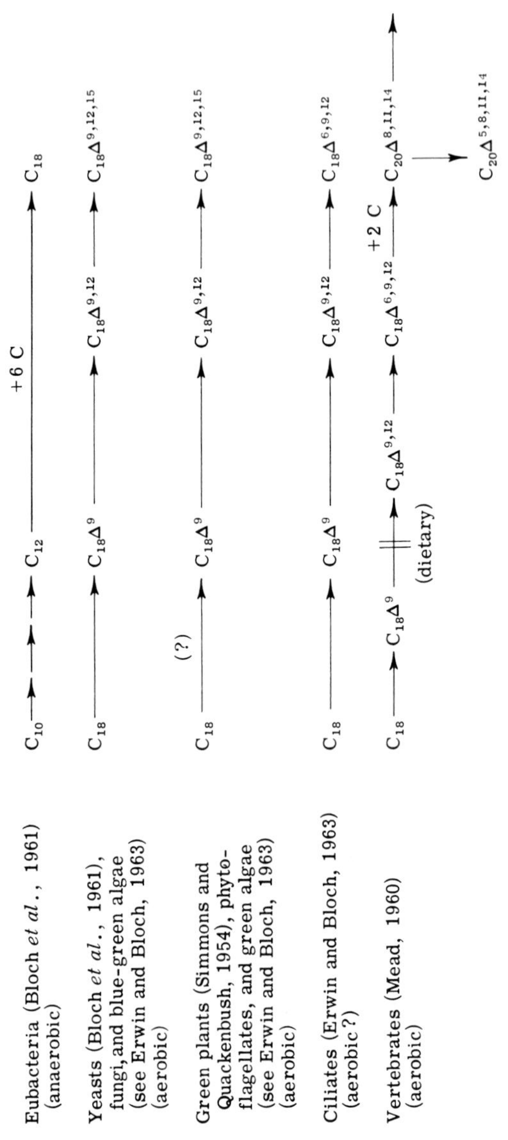

FIG. 3.1. Pathways of unsaturated fatty acid synthesis in various organisms. (From Erwin and Bloch, 1963.)

Bloch (1963) point out that it is only necessary to envision the appearance of enzymes for chain elongation of γ-linolenate followed by further desaturation to arachidonate and the loss of the ability to desaturate oleate to linoleate in order to "evolve" the vertebrate pattern from the ciliate pattern. This tends to support a theory that Metazoa were derived from flagellates by the way of ciliates and the acoel *Turbellaria* (Hanson, 1958).

The biosynthesis of phospholipids by *T. pyriformis* has received recent attention. Cell-free extracts catalyze the following reactions (Dennis and Kennedy, 1970)

$$\text{Phosphatidyl ethanolamine} + \text{L-serine} \underset{}{\overset{Ca^{2+}}{\rightleftharpoons}} \text{phosphatidyl serine} + \text{ethanolamine}$$

$$\text{Phosphatidyl serine} \rightarrow \text{phosphatidyl ethanolamine} + CO_2$$

The net result of these reactions is the decarboxylation of serine. The process of biosynthesis of phosphatidyl serine resembles that found in higher animals, and not that in bacteria. Michaelis constants for substrates in the first reaction are, for serine, 0.3 mM; for ethanolamine, 0.1 mM; and for calcium ions, 0.9 mM. The reaction is greatly stimulated by Trition X-100. The presence of the decarboxylase may explain why little phosphatidyl serine was found by several investigators (Thompson, 1967; Erwin and Bloch, 1963).

Phosphatidyl choline biosynthesis proceeds, as in other cells, by methylation, from *S*-adenosyl-L-methionine, of phosphatidyl ethanolamine (Smith and Law, 1970b). The pH optimum for the reaction is 8.5. The enzyme system accepts only phosphatidyl monomethylethanolamine as an exogenous substrate, but endogenously complete methylation of phosphatidyl ethanolamine is achieved. Another process for forming phosphatidyl choline involves preformed choline. The choline is presumably phosphorylated and converted to cytidine diphosphate (CDP)-choline, which is the substrate of the CDP-choline–diglyceride phosphocholinetransferase in the mitochondria (Smith and Law, 1970b). The enzyme has a pH optimum of 8 to 8.5 and requires manganese ions. The transferase is present in animals but is absent in the protozoans *Euglena* and *Ochromonas* (Lust and Daniel, 1966; Tipton and Swords, 1966).

Ethanolamine and choline phospholipids are readily degraded by extracts of *T. pyriformis* (Thompson, 1967), probably in part by the phospholipases which have been identified in particles from this organism (Eichel and Rem, 1963). The activity of the lipolytic enzymes increases with age (Thompson, 1969a); and, at maximum levels of activity, the enzymes can hydrolyze 60% of the endogenous phospholipids within 1 hr. A product of phospholipid degradation, choline, can be further metabolized, for choline oxidase is present (Allen, 1968).

Studies have been made on the metabolism by *T. pyriformis* of chimyl alcohol (1-*O*-hexadecylglycerol), a component of the glyceryl ethers (Kapoulas *et al.*, 1969a,b). Cells exposed to a low level incorporate most of it directly, without cleavage, into phospholipids. As the amount of administered substrate increases, the content of phospholipids containing glyceryl ethers reaches maximal levels; and other, degradative, pathways become operative. The increased ability to degrade chimyl alcohol is not due to the action of induced enzymes.

More than for intact cells, crude homogenates of the organism actively cleave chimyl alcohol (Kapoulas *et al.*, 1969a,b). The activity is associated with the microsomal fraction; and the reaction requires oxygen, NAD^+, and a reduced NADP regenerating system. In contrast to liver preparations, no tetrahydropteridine is required. The major product is an unidentified fatty acid, but traces of a fatty aldehyde and fatty alcohol can be demonstrated. The system will cleave glyceryl–vinyl ethers and glycol ethers, but not glyceryl–ether phosphatides. No changes in enzyme activity of homogenates are noted at different stages of culture growth.

Incorporation studies with ^3H– and ^{14}C-labeled glycerol were not consistent with the direct incorporation of glycerol into glyceryl ethers (Friedberg and Greene, 1968). The investigators suggested that an alkyl glycoside might be cleaved in the process of forming glyceryl ethers. However, a recent investigation shows that a cell-free preparation from *T. pyriformis* can utilize glyceraldehyde 3-phosphate for biosynthesis of the glyceryl moiety (Kapoulas and Thompson, 1969). Cofactors required are coenzyme A, adenosine triphosphate (ATP), and magnesium ions; and reduced nicotinamide adenine dinucleotide phosphate (NADP) greatly stimulates glyceryl ether synthesis.

Tetrahymanol and Sterol Metabolism

Some investigations on the biosynthesis of tetrahymanol have been made. Two brief reports state that tetrahymanol is labeled by acetate but not by squalene (Shorb *et al.*, 1966) and that mevalonate is incorporated into tetrahymanol (Holz and Erwin, 1958). More elaborate investigations, however, show that either acetate, mevalonate, or squalene can be incorporated (Conner *et al.*, 1968; Caspi *et al.*, 1968a,b). The formation of tetrahymanol and its companion compound is inhibited by the addition of cholesterol to the culture fluid. Cholesterol inhibits the biosynthesis of squalene from acetate and mevalonate and also the formation of tetrahymanol from squalene [reaction (3.1)]. The synthesis of other terpene derivatives (ubiquinones) from acetate and mevalonate continues in the presence of cholesterol, suggesting that a major block occurs after "isoprene" formation and before squalene formation (Conner *et al.*, 1968; 1969).

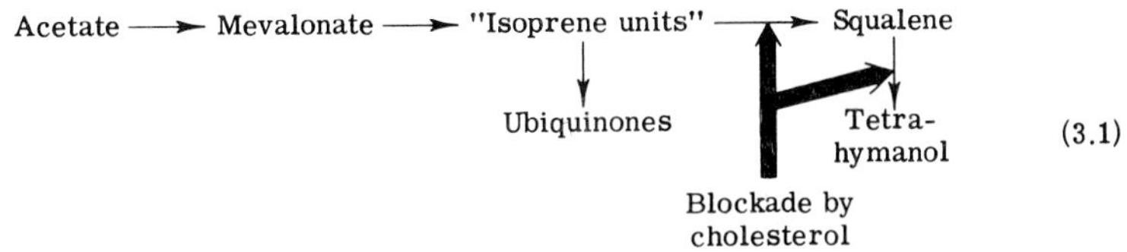

(3.1)

Anaerobic cyclization of squalene, followed by hydroxylation, gives rise to tetrahymanol without loss or gain of carbon units. Cyclization is induced by a proton rather than by an oxidant as in sterol synthesis since squalene, but not squalene 2,3-oxide, is incorporated into tetrahymanol by either intact cells or by broken cell preparations (Tsuda *et al.*, 1965; Caspi *et al.*, 1968a,b). A hydrogen atom is acquired from the medium, presumably at C-3; and an oxygen atom from water is added at C-21 (Zander *et al.*, 1970).

The metabolic basis for sterol requirements by *Tetrahymena corlissi*, *Tetrahymena paravorax*, and *Tetrahymena setifera* is not known. Perhaps the mechanism for forming tetrahymanol is blocked in the organisms requiring sterols, and the sterols which permit growth may have the necessary structure to fulfill the function of tetrahymanol. Another possibility, suggested by the structural similarity of the active compounds, is that all of them could be converted metabolically to the pentacyclic triterpenoid alcohol. One action of these sterols may be related to the biosynthesis of unsaturated fatty acids; *Tetrahymena setifera* requires cholesterol for the optimum formation of these compounds (Erwin *et al.*, 1966). In *Tetrahymena pyriformis*, tetrahymanol may fill the membrane structure role accomplished by sterols in other eukaryotic cells.

Although *T. pyriformis* does not require sterols, it can act metabolically on sterols supplied in the medium (Fig.3.2). Cholesterol is converted to cholesta-5,7—*trans*-22-trien-3β-ol (Conner and Ungar, 1964; Landrey *et al.*, 1968; Conner *et al.*, 1969). The isolation and identification of this triene can be achieved by using cultures grown in the presence of cholesterol (Mallory *et al.*, 1968). 7-Dehydrocholesterol and 22-dehydrocholesterol are also products, and it is not clear which is the intermediate leading to the triene, for both give rise to it. During the conversion, formation of the 7,8-double bond involves removal of the 7β- and 8β-hydrogen atoms (Wilton and Akhtar, 1970; Zander and Caspi, 1970). The 22-*pro*-R hydrogen is removed from C-22 (Zander and Caspi, 1970). Cholestanol, β-sitosterol, and stigmasterol are also metabolized to the corresponding trienes (Conner *et al.*, 1966). The dienes, ostreasterol and isofucosterol, are transformed to tetraenes; but the diene fucosterol is converted only

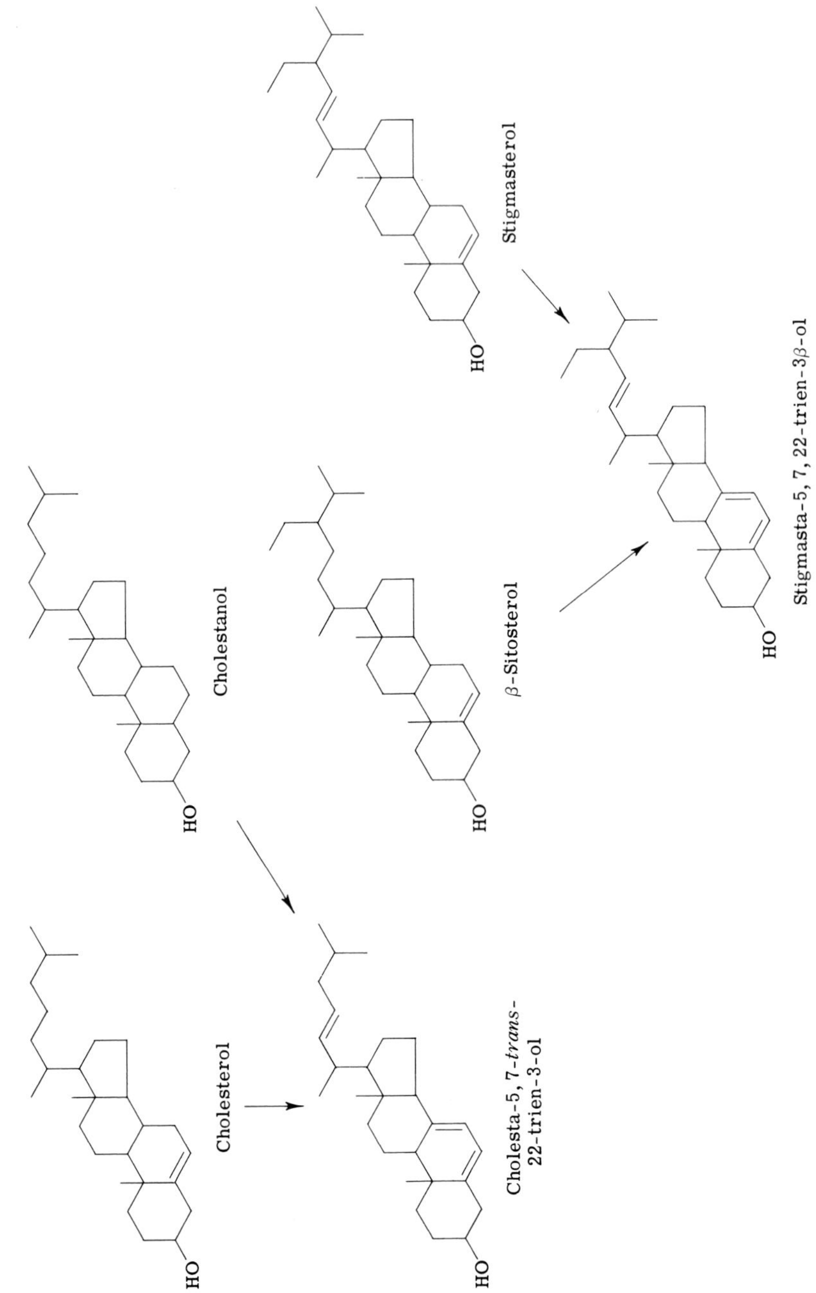

Stigmasterol

Stigmasta-5, 7, 22-trien-3β-ol

Cholestanol

β-Sitosterol

Cholesterol

Cholesta-5, 7-*trans*-22-trien-3-ol

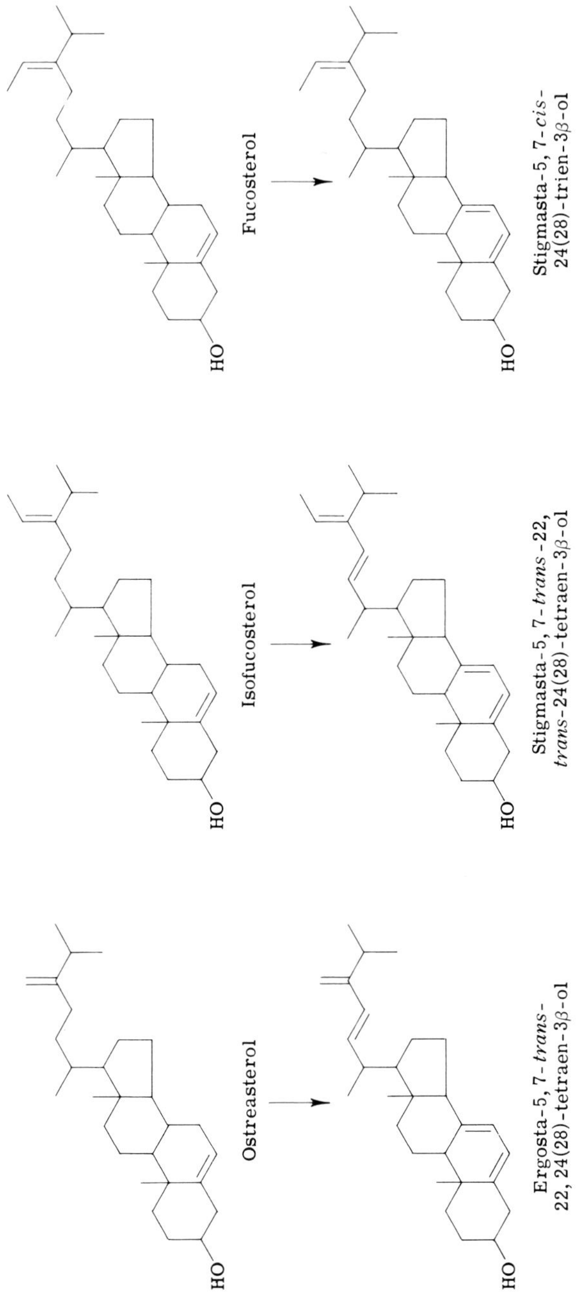

Ostreasterol → Ergosta-5, 7-*trans*-22, 24(28)-tetraen-3β-ol

Isofucosterol → Stigmasta-5, 7-*trans*-22, *trans*-24(28)-tetraen-3β-ol

Fucosterol → Stigmasta-5, 7-*cis*-24(28)-trien-3β-ol

Fig. 3.2. Metabolism of sterols by *Tetrahymena pyriformis*.

to a triene, with no double bond being added at C-22 (Nes *et al.*, 1971). The failure of the C-22 oxidoreductase to function on this compound is attributed to a steric effect. Lophenol, which can be utilized as a source of sterol for *T. corlissi*, can be recovered unchanged after growth of this organism (Britt and Bloch, 1961), thus supporting the hypothesis that *Tetrahymena* may utilize sterols without metabolic alteration.

An enzyme capable of deacylating β-hydroxy-β-methylglutaryl CoA, an intermediate in sterol biosynthesis, is active in *T. pyriformis* (Dekker *et al.*, 1958).

Not only do certain sterols inhibit the growth of *T. pyriformis*, but compounds which are known to impair the biosynthesis of cholesterol in other organisms are also effective in inhibiting growth and in blocking division of synchronized cells (Aaronson *et al.*, 1962; Aaronson and Bensky, 1965; Johnson and Jasmin, 1961; Holmlund, 1970; Holz *et al.*, 1963). One such agent, triparanol, has been studied in some detail. Its inhibits both growth in a defined medium and synchronous division in an inorganic medium. Its action on growth is prevented by certain sterols and/or long-chain fatty acids (Holz *et al.*, 1962b; Aaronson *et al.*, 1964; Pollard *et al.*, 1964). Synthetic phosphatides potentiate the action of sterols and fatty acids in anulling growth inhibition by triparanol. Inhibition of the first synchronous division is prevented by a variety of oleic acid-containing compounds, but not by sterols (Holz *et al.*, 1963). From the specificity of growth responses to sterols in the presence of triparanol, the investigators judge that triparanol impairs reduction of the 24 (25) double bond of sterols such as ostreasterol and fucosterol and of the 24 (28) double bond of sterols such as zymosterol and desmosterol during their conversion to ciliate polycyclic compounds.

From a study of the effects of triparanol in reducing lipid synthesis in *T. pyriformis*, Pollard *et al.* (1964) speculate that its action may be on the demethylation and oxidation of saturated fatty acids, as there is a reduction in the proportion of unsaturated to saturated acids. The same group notes that triparanol produces an increase in the amount of squalene, which becomes the major nonsaponifiable lipid, and a decrease in the amount of tetrahymanol in the cells (Shorb *et al.*, 1964; 1965).

Sterols have the ability to antagonize the growth inhibitory action of triparanol and also the inhibition caused by 6-methylpurine and a number of other purine analogs (Dewey *et al.*, 1959, 1960; Dewey and Kidder, 1962); 2,4-dinitrophenol (Conner, 1957; Conner *et al.*, 1961); puromycin (Dewey, 1967); 2,6-diamino-4-isoamyloxypyridine (Markees *et al.*, 1960); acetate antimetabolites (Dewey and Kidder, 1960); and colchicine, deoxycorticosterone, cortisone, and progesterone (Conner and Nakatani,

1958). The reversal of inhibition by 2,4-dinitrophenol is complicated by the fact that the composition of the medium influences the reversal effect (Conner, 1958). The presence of sterols enhances the uptake of phosphate from the medium, whereas 2,4-dinitrophenol depresses this uptake (Conner *et al.*, 1961). Cholesterol does not affect the uncoupling caused by 2,4-dinitrophenol; but it does prevent the loss of hypoxanthine, the end product of purine degradation, from the cells (Conner and Longobardi, 1963). Koroly and Conner (1966), investigating factors influencing ribonucleic acid metabolism, concluded that cholesterol either prevents the breakdown of nucleic acids or promotes their synthesis.

Other lipids are able to reverse the growth inhibition by a number of agents. Phosphatidyl choline (lecithin) reverses the effects of the antibiotics tyrocidine and puromycin (Dewey, 1967), the tranquilizer chlorpromazine (Guttman and Friedman, 1967; Dewey, 1967); and 2,6-diamino-4-isoamyloxypyridine (Markees *et al.*, 1960). Tween 80 overcomes the inhibition by tyrocidine, the detergent Triton X-100, chlorpromazine, and the uncoupling agent carbonylcyanide *m*-chlorophenylhydrazone (Dewey, 1967).

Cholinesterase

There has been some dispute as to whether *Tetrahymena pyriformis* possesses acetylcholine and acetylcholinesterase activity. Seaman and Houlihan (1951) reported that acetylcholine was split by this organism, and a brief communication by Straughn and Ronkin (1961) refers to the presence of acetylcholinesterase. However, the rate of cleavage, as reported by Seaman, is about 0.08 μg per mg of protein per hour, a rate within the experimental error of the assay (Nachmanson and Wilson, 1955). Other assays reveal that no acetylcholine and no acetylcholinesterase is present in this organism (Torres de Castro and Couceiro, 1955; Tibbs, 1960; Aaronson and Bensky, 1963). The evidence presented by the latter investigators is impressive in that enzyme assays and standard histochemical methods yield negative results and in that diisopropylfluorophosphate, eserine, and neostigmine are not toxic.

Although acetylcholine is not cleaved to a detectable extent, other esters of choline are broken down. Myristyl, lauryl, and benzoylcholine are substrates for esterases in the cell (Pastor and Fennell, 1959). Possibly the enzyme cleaving these substrates is the one localized in the pellicle, as seen by the lead thiolacetic acid method (Schuster and Hershenov, 1969).

Phosphonolipids

When first discovered, the phosphonic acids present in ciliates were the only compounds known in nature to contain a C—P bond. The structures of those present in *Tetrahymena pyriformis* are as follows:

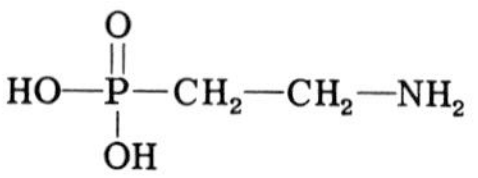

$$HO-\overset{\overset{O}{\|}}{\underset{\underset{OH}{|}}{P}}-CH_2-CH_2-NH_2$$

2-Aminoethylphosphonic acid (ciliatine)

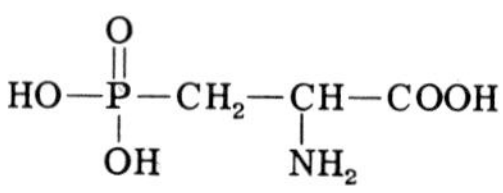

$$HO-\overset{\overset{O}{\|}}{\underset{\underset{OH}{|}}{P}}-CH_2-\underset{\underset{NH_2}{|}}{CH}-COOH$$

2-Amino-3-phosphonopropionic acid

The phosphonates are now known to be widely distributed. They have been found in six species of phytoplankton (Kittredge *et al.*, 1969; Baldwin and Braven, 1968), in two ciliates (Kandatsu and Horiguchi, 1962; Carter and Gaver, 1967), in twenty-five species of coelenterates (Quin, 1964, 1965; Kittredge, 1964), and in thirty-three species of mollusks (Quin, 1964, 1965; Simon and Rouser, 1967; Hori *et al.*, 1964a,b; 1966, 1967; Higashi and Hori, 1968). They have also been detected in nematodes, annelids, echinoderms, and Crustacea (Quin, 1964, 1965; Shimazu *et al.*, 1965). There is some evidence that the phosphonates occur in goat liver (Kandatsu and Horiguchi, 1965) and other mammalian tissues, but these may have been assimilated from the digestive tract. A review of biochemical carbon–phosphorus bonds has recently appeared (Kittredge and Roberts, 1969).

These compounds can be isolated from ^{32}P-labeled *T. pyriformis* (Kandatsu and Horiguchi, 1962; Rosenberg, 1964; Kittredge and Hughes, 1964; Warren, 1968); the ^{32}P is incorporated into these materials only during the logarithmic phase of growth (Rosenberg, 1964). In *T. pyriformis*, 2-aminoethylphosphonate occurs as the phosphonate analog of phosphatidyl ethanolamine (Liang and Rosenberg, 1966) and as glyceryl monoether phosphatidyl aminoethylphosphonate (Thompson, 1967). The phosphonic acid analog of phosphatidyl ethanolamine is only a minor component; the ratio of this compound to phosphatidyl ethanolamine is 1:13 (Liang and Rosenberg, 1966). The phosphonolipids resist attack by the endogenous lipolytic enzymes of *T. pyriformis* (Thompson, 1969b), and such resistance may confer stability to the membranes of which these molecules are a part (Kittredge and Roberts, 1969; Kennedy and Thompson, 1970). Surface membrane fractions of *T. pyriformis*, particularly membranes surrounding cilia, contain higher percentages of phosphonolipids than do other membranes of the cell (Kennedy and Thompson, 1970). Lipid extracts of isolated cilia of *T. pyriformis* on thin-layer chromatography show spots corresponding to phosphatidyl inositides, phosphatidyl ethanolamine, and phosphatidyl choline (Culbertson and Hull, 1963); but by chemical analysis, the major polar component is a glyceryl ester of 2-aminoethylphosphonic acid (Smith *et al.*, 1970; Jonah and Erwin, 1971). Thin-layer chromatography is not adequate for characterization of phosphonolipids. Perhaps a newly developed assay involving ^{31}P-nuclear magnetic resonance will be useful (Glonek *et al.*, 1970). The glyceryl ester of 2-aminoethylphosphonic acid is also found in mitochondria, along with phosphatidyl ethanolamine, phosphatidyl choline, and cardiolipin (Jonah and Erwin, 1971).

2-Amino-3-phosphonopropionate reportedly occurs in polypeptide linkage rather than lipid (Kittredge and Hughes, 1964). Compounds that contain "sugar," glycerol, ethanolamine, and 2-aminoethylphosphonate accumulate in heat-treated cells and are possibly involved in cell division (Chou and Scherbaum, 1965). What relationship these unusual substances have to the phosphonate analogs of phosphatides is not known.

Tetrahymena pyriformis have been used to study the biosynthesis of the phosphonate analog of phosphatidyl ethanolamine (Liang and Rosenberg, 1966). 2-Aminoethylphosphonate, in the presence of cytidine triphosphate (CTP), is converted to a nucleotide-bound form identified as cytidine monophosphate (CMP)-aminoethylphosphonate. This product subsequently reacts with a diglyceride to form the phosphonate-containing glycerophosphatide. This pathway apparently is concerned with reutilization of 2-aminoethylphosphonate. In the biosynthesis of this compound, lipid-bound 2-aminoethylphosphonate picks up the label from ^{32}P much more rapidly than soluble 2-aminoethylphosphonate (Rosenberg, 1964). Further support for a lipid being involved comes from Thompson (1969b), who found that the incorporation of labeled palmitate and chimyl alcohol into choline phospholipids of *T. pyriformis* is followed by a gradual transfer of the labeled glyceride moiety from that fraction to 2-aminoethylphosphonate or to a precursor of this compound.

For the carbon skeleton, glucose (1- or 6-^{14}C-labeled) is a good precursor (Trebst and Geike, 1967; Warren, 1968; Liang and Rosenberg, 1968), being considerably better than pyruvate, acetate, alanine, glutamate, aspartate, serine, or ethanolamine (Liang and Rosenberg, 1968; Horiguchi *et al.*, 1968). It is noteworthy that serine and ethanolamine are poor precursors of the carbon skeleton, although they rapidly label lipid ethanolamine. The suggested biosynthesis of 2-aminoethylphosphonate from phosphatidyl ethanolamine (Segal, 1965) is not supported by these data.

The rapid incorporation of glucose is compatible with the assumption that the precursor is one of the glycolytic intermediates, of which phosphoenolpyruvate is most likely. This compound may react with an ester of phosphatidic acid to yield a glyceride ester of phosphoenolpyruvate which, on rearrangement, transamination, and decarboxylation, would yield a phosphonolipid (Liang and Rosenberg, 1968).

Warren (1968) suggests an intramolecular rearrangement of phosphoenolpyruvate, where both the phosphorus and the carbon are incorporated into 2-aminoethylphosphonate. This scheme involves free 2-aminoethylphosphonate as a precursor of the lipid-bound compound; but this is not compatible with the labeling pattern obtained with ^{32}P (Rosenberg, 1964).

Investigators using ^{14}C-labeled substrates note that 2-aminoethylphosphonic acid is incorporated into phospholipids without any degradation (Smith and Law, 1970a). A sizable portion of 2-amino-3-phosphonopropionic acid is found as 2-aminoethylphosphonic acid in the phospholipids, but no 2-amino-3-phosphonopropionic acid is in this fraction (Smith and Law, 1970a).

Broken cell preparations decarboxylate 2-amino-3-phosphonopropionate to form 2-aminoethylphosphonate (Warren, 1968), and one can use dialyzed homogenates of *T. pyriformis* and α-ketoglutarate to transaminate 2-aminoethylphosphonate and 2-amino-3-phosphonopropionate (Roberts *et al.*, 1968). The following analogs are also transaminated: 3-aminopropylphosphonate, 2-amino-4-phosphonobutyrate, DL-1,2-diaminoethylphosphonate, and aminomethylphosphonate. The metabolic significance of decarboxylation and transamination is not clear, for this is likely not to be the mechanism for formation of 2-aminoethylphosphonate.

Synthetic Lipids as Growth Factors

A chapter on the lipids of *Tetrahymena* would be incomplete without mention of the synthetic lipids Tween 80 (polyoxyethylene ester of sorbitan monooleate), TEM-4T (diacetyl tartaric acid esters of tallow monoglycerides), and Myrj G2144 (polyoxyalkylene derivative of oleic acid), which have been used in the culture of this ciliate (Kidder and Dewey, 1951; Holz *et al.*, 1961b; Miller and Johnson, 1960; Epstein *et al.*, 1963). These substances serve as a soluble (dispersible), relatively nontoxic, source of fatty acids. They have been used with the understanding that they stimulate growth and lead to a larger yield of organisms. Nevertheless, studies have shown that, in fact, growth is not stimulated but that uptake of these lipids leads to increased coloration of the cells, which accounts for the increased absorbance in turbidimetric measurement of growth (Kidder *et al.*, 1954; Jasmin, 1961). Fragility, which is undesirable for many experimental applications, is also noted when cells are grown in the presence of these synthetic lipids. In view of this evidence, use of Tween 80, TEM-4T, or Myrj G2144 in media for growth of *Tetrahymena* can no longer be recommended for all cultures. Tween 80 is still useful for applications where fragility of the cells is not important.

REFERENCES

Aaronson, S., and Baker, H. (1961). *J. Protozool.* **8,** 274.

Aaronson, S., and Bensky, B. (1963). *J. Protozool.* **10,** Suppl. 8.

Aaronson, S., and Bensky, B. (1965). *J. Protozool.* **12,** 236.

Aaronson, S., Bensky, B., Shifrine, M., and Baker, H. (1962). *Proc. Soc. Exp. Biol. Med.* **109,** 130.

Aaronson, S., Bensky, B., Haines, T. H., Gellerman, J. L., and Schlenk, H. (1963). *J. Protozool.* **10,** Suppl. 9.

Aaronson, S., Baker, H., Bensky, B., Frank, O., and Zahalsky, A. C. (1964). *Develop. Ind. Microbiol.* **6,** 48.

Allen, S. L. (1968). *Ann. N.Y. Acad. Sci.* **151,** 190.

Allison, B. M., and Ronkin, R. R. (1967). *J. Protozool.* **14,** 313.

Avins, L. R. (1968). *Biochem. Biophys. Res. Commun.* **32,** 138.

Baldwin, M. W., and Braven, J. (1968). *J. Mar. Biol. Ass. U.K.* **48,** 603.

Bloch, K., Barnowsky, P., Goldfine, H., Lennarz, W. J., Light, R., Norris, A. T., and Scheurerbrandt, G., (1961). *Fed. Proc. Fed. Amer. Soc. Exp. Biol.* **20,** 921.

Blum, J. J., and Wexler, J. P. (1968). *Mol. Pharmacol.* **4,** 155.

Britt, J. J., and Bloch, K. (1961). *Comp. Biochem. Physiol.* **2,** 213.

Carter, H. E., and Gaver, R. C. (1967). *Biochem. Biophys. Res. Commun.* **29,** 886.

Castrejon, R. N. (1964). *Diss. Abstr.* **24,** 3083.

Caspi, E., Zander, J. M., Greig, J. B., Mallory, F. B., Connor, R. L., and Landrey, J. R. (1968a). *J. Amer. Chem. Soc.* **90,** 3563.

Caspi, E., Greig, J. B., and Zander, J. M. (1968b). *Biochem. J.* **109,** 931.

Chaix, P., and Baud, C. (1947). *Arch. Sci. Physiol.* **1,** 3.

Chaix, P., Chauvet, J., and Fromageot, C. (1947). *Antonie van Leeuwenhoek J. Microbiol. Serol.* **12,** 145.

Chou, S. C., and Scherbaum, O. H. (1965). *Exp. Cell Res.* **45,** 31.

Conger, N. E., and Eichel, H. J. (1965). *Fed. Proc. Fed. Amer. Soc. Exp. Biol.* **24,** 350.

Conner, R. L. (1957). *Science* **126,** 698.

Conner, R. L. (1958). *J. Protozool.* **5,** Suppl. 25.

Conner, R. L. (1959). *J. Gen. Microbiol.* **21,** 180.

Conner, R. L., and Longobardi, A. E. (1963). *J. Protozool.* **10,** Suppl. 8.

Conner, R. L., and Nakatani, M. (1958). *Arch. Biochem. Biophys.* **74,** 175.

Conner, R. L., and Ungar, F. (1964). *Exp. Cell Res.* **36,** 134.

Conner, R. L., Kornacker, M. S., and Goldberg, R. (1961). *J. Gen. Microbiol.* **26,** 437.

Conner, R. L., Iyengar, C. W. L., and Landrey, J. R. (1966). *J. Protozool.* **13,** Suppl. 13.

Conner, R. L., Landrey, J. R., Burns, C. H., and Mallory, F. B. (1968). *J. Protozool.* **15,** 600.

Conner, R. L., Mallory, F. B., Landrey, J. R., and Iyengar, C. W. L. (1969). *J. Biol. Chem.* **244,** 2325.
Cox, D. (1970). *J. Protozool.* **17,** 150.
Culbertson, J. R., and Hull, R. W. (1963). *J. Protozool.* **10,** Suppl. 8.
Dekker, E. E., Schlesinger, M. J., and Coon, M. J. (1958). *J. Biol. Chem.* **233,** 434.
Den, H., Robinson, W. G., and Coon, M. J. (1959). *J. Biol. Chem.* **234,** 1666.
Dennis, E. A., and Kennedy, E. P. (1970). *J. Lipid Res.* **11,** 394.
Dewey, V. C. (1967). *In* "Chemical Zoology" (M. Florkin and B. T. Scheer, eds.), Vol. 1, p. 162. Academic Press, New York.
Dewey, V. C., and Kidder, G. W. (1960). *Arch. Biochem. Biophys.* **88,** 78.
Dewey, V. C., and Kidder, G. W. (1962). *Biochem. Pharmacol.* **11,** 53.
Dewey, V. C., Kidder, G. W., and Markees, D. G. (1959). *Proc. Soc. Exp. Biol. Med.* **102,** 306.
Dewey, V. C., Heinrich, M. R., Markees, D. G., and Kidder, G. W. (1960). *Biochem. Pharmacol.* **3,** 173.
Eichel, H. J., and Rem, L. T. (1963). *In* "Progress in Protozoology" (J. Ludvik *et al.*, eds.), p. 148. Academic Press, New York.
Engermann, J. G. (1958). *J. Protozool.* **5,** Suppl. 13.
Epstein, S. S., Burroughs, M., and Small, M. (1963). *Cancer Res.* **23,** 35.
Erwin, J. A. (1970). *Biochim. Biophys. Acta* **202,** 21.
Erwin, J., and Bloch, K. (1963). *J. Biol. Chem.* **238,** 1618.
Erwin, J. A., Beach, D., and Holz, G. G., Jr. (1966). *Biochim. Biophys. Acta* **125,** 614.
Everhart, L. P., and Ronkin, R. R. (1966). *J. Protozool.* **13,** 646.
Friedberg, S. J., and Greene, R. C. (1968). *Biochim. Biophys. Acta* **164,** 602.
Glonek, T., Henderson, T. O., Hilderbrand, R. L., and Myers, T. C. (1970). *Science* **169,** 192.
Gordon, J. T., and Doyne, T. H. (1966). *Acta Cryst.* **21,** Suppl. A113.
Green, J., Price, S. A., and Gare, L. (1959). *Nature (London)* **184,** 1339.
Guttman, H. N., and Friedman, W. (1967). *Trans. N.Y. Acad. Sci.* **26,** 75.
Hack, M. H., Yeager, R. G., and McCaffery, T. D. (1962). *Comp. Biochem. Physiol.* **6,** 247.
Haines, T. H. (1965). *J. Protozool.* **12,** 655.
Hanson, C. D. (1958). *Syst. Zool.* **7,** 16.
Helmer, E. (1968). *J. Protozool.* **15,** Suppl. 36.
Higashi, S., and Hori, T. (1968). *Biochim. Biophys. Acta* **152,** 568.
Holmlund, C. E. (1970). *J. Protozool.* **17,** Suppl. 12.
Holmlund, C. E., and Bohonos, N. (1966). *Life Sci.* **5,** 2133.
Holz, G. G., Jr. (1963). *In* "Progress in Protozoology" (J. Ludvik *et al., eds.*), p. 222. Academic Press, New York.
Holz, G. G., Jr., and Erwin, J. (1958). *J. Protozool.* **5,** Suppl. 19.
Holz, G. G., Jr., Wagner, B., Erwin, J., Britt, J. J., and Bloch, K. (1961a). *Comp. Biochem. Physiol.* **2,** 202.
Holz, G. G., Jr., Erwin, J. A., and Wagner, B. (1961b). *J. Protozool.* **8,** 297.
Holz, G. G., Jr., Erwin, J., Wagner, B., and Rosenbaum, N. (1962a). *J. Protozool.* **9,** 359.
Holz, G. G., Jr., Erwin, J., Rosenbaum, N., and Aaronson, S. (1962b). *Arch. Biochem. Biophys.* **98,** 312.
Holz, G. G., Jr., Rasmussen, L., and Zeuthen, E. (1963). *C. R. Trav. Lab. Carlsberg* **33,** 289.

Hori, T., Itasaka, O., Hoshimoto, T., and Inoue, H. (1964a). *J. Biochem.* (*Tokyo*) **55,** 545.
Hori, T., Itasaka, O., Inoue, H., and Yamada, K. (1964b). *J. Biochem.* (*Tokyo*) **56,** 477.
Hori, T., Itasaka, O., and Inoue, H. (1966). *J. Biochem.* (*Tokyo*) **59,** 570.
Hori, T., Arakawa, I., and Sugita, M. (1967). *J. Biochem.* (*Tokyo*) **62,** 67.
Horiguchi, H., Kittredge, J. S., and Roberts, E. (1968). *Biochim. Biophys. Acta* **165,** 164.
Hutner, S. H., and Holz, G. G., Jr. (1962). *Annu. Rev. Microbiol.* **16,** 189.
Jasmin, R. (1961). *Rev. Can. Biol.* **20,** 813.
Johnson, W. J., and Jasmin, R. (1961). *Fed. Proc. Fed. Amer. Soc. Exp. Biol.* **20,** 281.
Jonah, M., and Erwin, J. A. (1971). *Biochim. Biophys. Acta* **231,** 80.
Kandatsu, M., and Horiguchi, M. (1962). *Agr. Biol. Chem.* **26,** 721.
Kandatsu, M., and Horiguchi, M. (1965). *Agr. Biol. Chem.* **29,** 781.
Kapoulas, V. M., and Thompson, G. A., Jr. (1969). *Biochim. Biophys. Acta* **187,** 594.
Kapoulas, V. M., Thompson, G. A., Jr., and Hanahan, D. J. (1969a). *Biochim. Biophys. Acta* **176,** 237.
Kapoulas, V. M., Thompson, G. A., Jr., and Hanahan, D. J. (1969b). *Biochim. Biophys. Acta* **176,** 250.
Kennedy, K. E., and Thompson, G. A., Jr. (1970). *Science* **168,** 989.
Kessler, D. (1961). M.S. thesis, Syracuse University.
Kidder, G. W., and Dewey, V. C. (1949). *Arch. Biochem.* **20,** 433.
Kidder, G. W., and Dewey, V. C. (1951). *In* "Biochemistry and Physiology of Protozoa" (A. Lwoff, ed.), Vol. 1, p. 323. Academic Press, New York.
Kidder, G. W., Dewey, V. C., and Heinrich, M. R. (1954). *Exp. Cell Res.* **7,** 256.
Kittredge, J. S. (1964). M. S. thesis, University of California.
Kittredge, J. S., and Hughes, R. R. (1964). *Biochemistry* **3,** 991.
Kittredge, J. S., and Roberts, E. (1969). *Science* **164,** 37.
Kittredge, J. S., Williams, P. M., and Horiguchi, M. (1969). *Comp. Biochem. Physiol.* **29,** 859.
Knuese, W. R., and Shorb, M. S. (1966). *J. Protozool.* **13,** Suppl. 12.
Koehler, L. D., and Fennell, R. A. (1964). *J. Morphol.* **114,** 209.
Koroly, M. J., and Conner, R. L. (1966). *J. Protozool.* **13,** Suppl. 13.
Kozak, M. S., and Huddleston, M. S. (1966). *J. Protozool.* **13,** Suppl. 27.
Landrey, J. R., Conner, R. L., and Iyengar, C. W. L. (1968). *J. Protozool.* **15,** Suppl. 14.
Lees, A. M., and Korn, E. D. (1966). *Biochemistry* **5,** 1475.
Levy, M. R., and Elliott, A. M. (1968). *J. Protozool.* **15,** 208.
Liang, C. R., and Rosenberg, H. (1966). *Biochim. Biophys. Acta* **125,** 548.
Liang, C. R., and Rosenberg, H. (1968). *Biochim. Biophys. Acta* **156,** 437.
Lust, G., and Daniel, L. J. (1966). *Arch. Biochem. Biophys.* **113,** 603.
McKee, C. M., Dutcher, J. D., Groupé, V., and Moore, M. (1947). *Proc. Soc. Exp. Biol. Med.* **65,** 326.
Mallory, F. B., Gordon, J. T., and Conner, R. L. (1963). *J. Amer. Chem. Soc.* **85,** 1362.
Mallory, F. B., Conner, R. L., Landrey, J. R., and Iyengar, C. W. L. (1968). *Tetrahedron Lett.*, p. 6103.
Markees, D. G., Dewey, V. C., and Kidder, G. W. (1960). *Arch. Biochem. Biophys.* **86,** 179.
Mead, J. F. (1960). *In* "Lipide Metabolism" (K. Bloch, ed.), p. 41. Wiley, New York.
Miller, C. A., and Johnson, W. H. (1960). *J. Protozool.* **7,** 297.
Müller, M., Farkas, T., and Herodek, S. (1959). *J. Protozool.* **6,** Suppl. 28.

Nachmanson, D., and Wilson, I. B. (1955). *In* "Methods in Enzymology" (S. P. Colowick and N. O. Kaplan, eds.), Vol. 1, p. 642. Academic Press, New York.
Nes, W. R., Malya, P. A. G., Mallory, F. B., Ferguson, K. A., Landrey, J. R., and Conner, R. L. (1971). *J. Biol. Chem.* **246,** 561.
Pace, D. M., and Ireland, R. L. (1945). *J. Gen. Physiol.* **28,** 547.
Pastor, E. P., and Fennell, R. A. (1959). *J. Morphol.* **104,** 143.
Pollard, W. O., Shorb, M. S., Lund, P. G., and Vasaitis, V. (1964). *Proc. Soc. Exp. Biol. Med.* **116,** 539.
Quin, L. D. (1964). *Science* **144,** 1133.
Quin, L. D. (1965). *Biochemistry* **4,** 324.
Reid, R., and Cox, D. (1968). *J. Protozool.* **15,** Suppl. 16.
Reid, R., Cox, D., Baker, H., and Frank, O. (1969). *J. Protozool.* **16,** 231.
Roberts, E., Simonsen, D. G., Horiguchi, M., and Kittredge, J. S. (1968). *Science* **159,** 886.
Rogers, C. G. (1968). *Can. J. Biochem.* **46,** 331.
Rosenbaum, N., Erwin, J., Beach, D., and Holz, G. G., Jr. (1966). *J. Protozool.* **13,** 535.
Rosenberg, H. (1964). *Nature (London)* **203,** 299.
Ryley, J. F. (1952). *Biochem. J.* **52,** 483.
Schneiderman, H. A., and Gilbert, L. I. (1964). *Science* **143,** 325.
Schneiderman, H. A., Gilbert, L. I., and Weinstein, M. (1960). *Nature (London)* **188,** 1041.
Schuster, F. L., and Hershenov, B. (1969). *Exp. Cell Res.* **55,** 385.
Seaman, G. R. (1950a). *J. Cell. Comp. Physiol.* **36,** 129.
Seaman, G. R. (1950b). *J. Biol. Chem.* **186,** 97.
Seaman, G. R., and Houlihan, R. K. (1951). *J. Cell. Comp. Physiol.* **37,** 309.
Segal, W. (1965). *Nature (London)* **208,** 1284.
Shaw, R. F., and Williams, N. E. (1963). *J. Protozool.* **10,** 486.
Shimazu, H., Kakimoto, Y., Nakajima, T., Kanasawa, A., and Sano, I. (1965). *Nature (London)* **207,** 1197.
Shorb, M. S. (1963). *In* "Progress in Protozoology" (J. Ludvik *et al.*, eds.), p. 153. Academic Press, New York.
Shorb, M. S., Dunlap, B., and Pollard, W. O. (1964). *J. Protozool.* **11,** Suppl. 25.
Shorb, M. S., Dunlap, B., and Pollard, W. O. (1965). *Proc. Soc. Exp. Biol. Med.* **118,** 1140.
Shorb, M. S., Knuese, W. R., and Pollard, W. O. (1966). *J. Protozool.* **13,** Suppl. 12.
Simmons, R. O., and Quackenbush, F. W. (1954). *J. Amer. Oil Chem. Soc.* **31,** 441.
Simon, G., and Rouser, G. (1967). *Lipids* **2,** 55.
Smith, J. D., and Law, J. H. (1970a). *Biochim. Biophys. Acta* **202,** 141.
Smith, J. D., and Law, J. H. (1970b). *Biochemistry* **9,** 2152.
Smith, J. D., Snyder, W. R., and Law, J. H. (1970). *Biochem. Biophys. Res. Commun.* **39,** 1163.
Straughn, M. C., and Ronkin, R. R. (1961). *Fed. Proc. Fed. Amer. Soc. Exp. Biol.* **20,** 134.
Taketomi, T. (1961). *Z. Allg. Mikrobiol.* **1,** 331.
Thompson, G. A., Jr. (1967). *Biochemistry* **6,** 2015.
Thompson, G. A., Jr. (1969a). *J. Protozool.* **16,** 397.
Thompson, G. A., Jr. (1969b). *Biochim. Biophys. Acta* **176,** 330.
Tibbs, J. (1960). *Biochim. Biophys. Acta* **41,** 115.
Tipton, C. L., and Swords, M. D. (1966). *J. Protozool.* **13,** 469.

Torres de Castro, F., and Couceiro, A. (1955). *Exp. Cell Res.* **8,** 245.
Trebst, A., and Geike, F. (1967). *Z. Naturforsch.* **B22,** 989.
Tsuda, Y., Morimoto, A., Sano, T., Inubushi, Y., Mallory, F. B., and Gordon, J. T. (1965). *Tetrahedron Lett.* p. 1427.
Vakirtzi-Lemonias, C., Kidder, G. W., and Dewey, V. C. (1963). *Comp. Biochem. Physiol.* **8,** 336.
Wagner, C. (1956). Ph.D. thesis, University of Michigan.
Warren, W. A. (1968). *Biochim. Biophys. Acta* **156,** 340.
Williams, B. L., Goodwin, T. W., and Ryley, J. F. (1966). *J. Protozool.* **13,** 227.
Wilton, D. C., and Akhtar, M. (1970). *Biochem. J.* **116,** 337.
Yuan, C., and Bloch, K. (1961). *J. Biol. Chem.* **236,** 1277.
Zander, J. M., and Caspi, E. (1970). *J. Biol. Chem.* **245,** 1682.
Zander, J. M., Caspi, E., Pandey, G. N., and Mitra, C. R. (1969). *Phytochemistry* **8,** 1597.
Zander, J. M., Greig, J. B., and Caspi, E. (1970). *J. Biol. Chem.* **245,** 1247.

CHAPTER 4

Energy Metabolism

Mitochondria

Mitochondria are fundamental to the energy metabolism of cells. Within these small, intracellular bodies are localized the enzymes of the Krebs cycle, the electron transport system which delivers electrons from the Krebs cycle intermediates to oxygen, and the obscure mechanism by which energy of this process is conserved by linking it to the formation of adenosine triphosphate (ATP). Structurally, the mitochondria of *Tetrahymena* (Fig. 4.1) are similar to those of other Protozoa. They have a double membrane and microvilli, which are the dominant internal features (Sedar and Rudzinska, 1956). A newly formed cell contains 600–800 mitochondria, and this number doubles before the next division (Sato, 1960). These particles apparently proliferate by self-division (Parsons, 1964). For cultures in exponential growth, most mitochondria are found near the kinetosomes (Fig. 4.2), which are the basal bodies for cilia (Elliott and Bak, 1964). In undisturbed, stationary-phase cells the number of mitochondria increases and they become rounder; but in stirred, stationary-phase cultures there

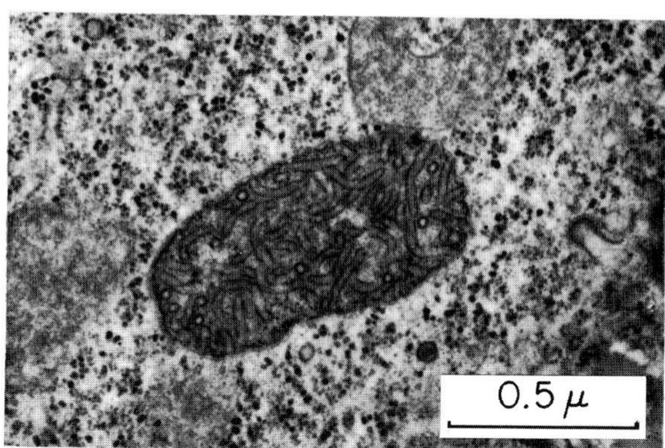

FIG. 4.1. Mitochondrion with typical microtubular system. Magnification: ×45,000. (Courtesy of Dr. I. L. Cameron and Mr. Glenn Williams.)

are relatively few mitochondria (Elliott *et al.*, 1966). During the early stages of starvation, there is degeneration of these structures, and this process is not prevented by the addition to the medium of an energy source such as glucose or acetate (Levy and Elliott, 1968). Some of the mitochondria are incorporated into vacuoles, where they are destroyed (Elliott and Bak, 1964).

The usual methods for preparation of mitochondria are not adequate for *Tetrahymena*, but suitable procedures involving gentle homogenization and differential centrifugation have been developed (Byfield *et al.*, 1962; Kobayashi, 1965). The median buoyant density of mitochondria is 1.211 (Brightwell *et al.*, 1968).

Respiration

Oxygen is required for growth of *Tetrahymena* (Pace and Ireland, 1945), but cells can survive in a static condition for several days under anaerobic conditions (Ryley, 1952). For *T. pyriformis* at 25°C in the absence of exogenous substrate, the rate of oxygen consumption ranges from 10 to 40 μl/mg dry weight/hr (Baker and Baumberger, 1941; McCashland and Kronschnabel, 1962; Ormsbee, 1942; Pace and Lyman, 1947; Danforth, 1967). Raising the temperature to 37°C approximately doubles the rate (Ryley, 1952). The rate also varies with the age of the culture and the concentration of cells. At high concentrations of cells, the reduced respiration is thought to be due less to a lack of oxygen than to CO_2 accumulation and

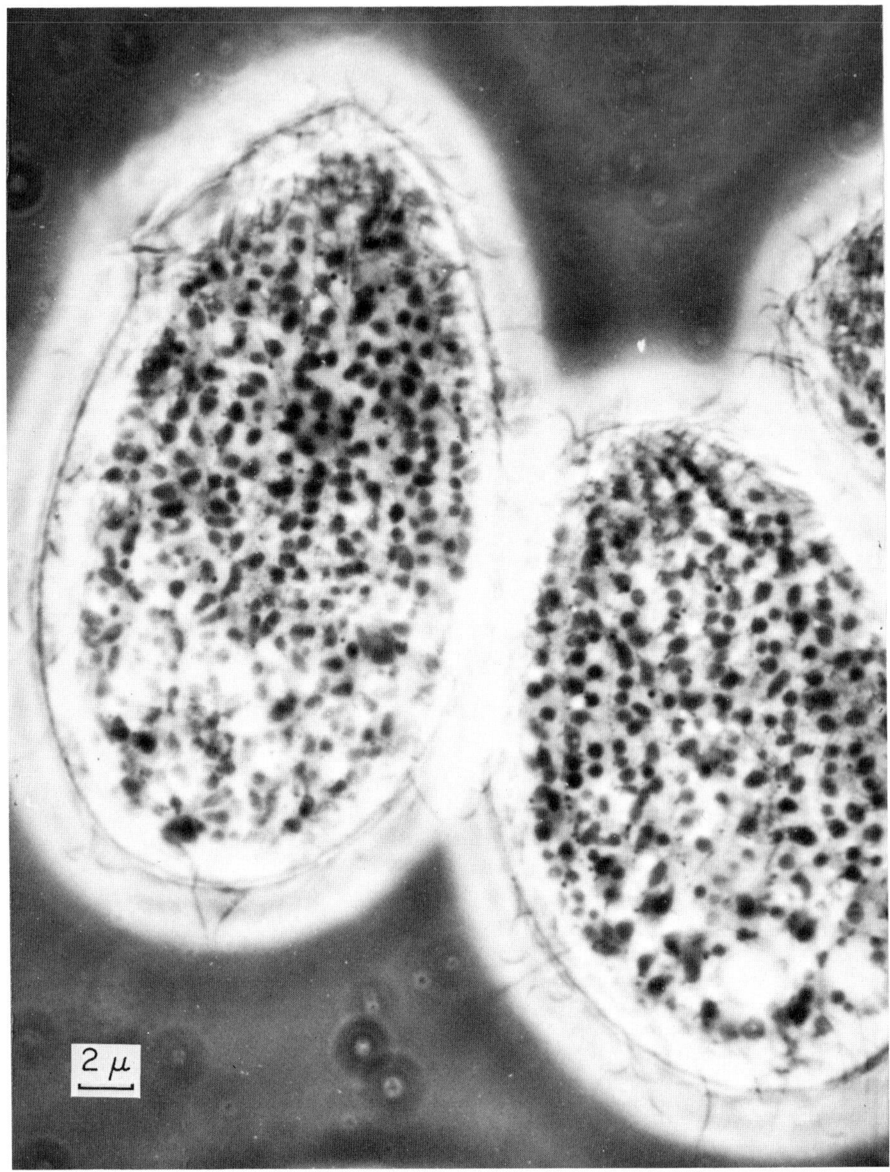

FIG. 4.2. Phase-contrast photograph showing mitochondria near kinetosomes. Magnification: ×3500. (Courtesy of Dr. I. L. Cameron and Mr. Glenn Williams.)

competition for nutrients (McCashland and Kronschnabel, 1962). High levels of CO_2 inhibit growth and respiration (Elliott *et al.*, 1962). One source (Ryley, 1952) states that the rate of endogenous metabolism remains practically unchanged for 5 hr in the absence of substrate, whereas in another publication it is reported that the rate decreases to 30% after 2 hr (Hamburger and Zeuthen, 1957).

Evidently associated with the increased energy requirement for the process of division, the rate of oxygen uptake for synchronized *T. pyriformis* just before division is more than threefold greater than that for exponentially growing cells (Hamburger and Zeuthen, 1957). Even short exposures to anaerobiosis cause a delay in the onset of division (Rasmussen, 1963). Related to this is the fact that the partial pressure of oxygen in the medium has an influence on the nucleoside triphosphate level in the cells. Cultures in which the O_2 tension is low and cell multiplication has almost ceased have nucleoside triphosphate levels almost twice as high as in exponentially growing cells for which ample O_2 is available (Scherbaum *et al.*, 1962).

Various organic substrates stimulate the respiration of *T. pyriformis.* Acetate, butyrate, and propionate almost double the rate (Ryley, 1952); and ethanol, isobutyrate, caprylate, palmitate, homocysteine, and methionine also increase the consumption of O_2 (Chaix *et al.*, 1947). Of a number of amino acids tested in another study, several stimulate respiration by amounts ranging from 10 to 50% above the endogenous rate (Roth *et al.*, 1954). The common amino acids causing stimulation are L-phenylalanine, L-tyrosine, D- and L-cysteine, L-proline, L-isoleucine, and D- and L-leucine. There is no adequate explanation for the failure of alanine, asparagine, glutamate, and glutamine to increase respiration of intact cells. Some dipeptides containing phenylalanine are more effective than the free amino acid (Roth and Eichel, 1961). Since the dipeptides are hydrolyzed outside the cell, their stimulation cannot be explained as a permeability phenomenon. The peptone medium, which is used for growth and which contains, among other substances, amino acids and peptides, stimulates respiration more than twofold (Ormsbee, 1942; Pace and Lyman, 1947; Hamburger and Zeuthen, 1957; van de Vijver, 1966a).

Cells that have been grown in the presence of glucose have higher respiration in the presence of glucose than cells that have not previously been exposed to this substance (Conner and Cline, 1967). Apparently, cells can adapt to various media by changing their level of O_2 uptake.

Ordinarily, glucose has only a relatively small effect on respiration; and the utilization of endogenous glycogen cannot account for the O_2 consumption of cells (Ryley, 1952). These data plus the facts that the lower fatty acids are used readily and that the respiratory quotient is 0.85 strongly suggest that lipids are the major substrates of aerobic metabolism.

Exposure to 600,000 r of X radiation reduces the O_2 consumption by 60 to 90% (Roth, 1962; Roth and Eichel, 1955). The decrease is prevented by adding 2,6-diaminopurine, pyruvate, or acetate during or after irradiation. γ-Irradiation also decreases respiration (van de Vijver, 1967).

The respiration of *T. pyriformis* is lowered by urethan (Lwoff, 1938; Eiler *et al.*, 1959) or phenylurethan (Bhagavan and Eiler, 1961). However, evidence indicates that the toxic effect of phenylurethan is not due to its effect on respiration.

Enzymes Associated with Electron Transport

As stated in Chapter 2, all of the enzymes of the Krebs cycle are present in *Tetrahymena pyriformis*. Isocitrate, α-ketoglutarate, succinate, and malate provide electrons for the reduction of oxygen (Eichel, 1959a; Kobayashi, 1965; Müller *et al.*, 1968). Concerning substrates from other sources, β-hydroxybutyrate, a product of fatty acid degradation, is effective, as are ascorbate (Kobayashi, 1965) and glutamate (Roth and Eichel, 1955); but α-glycerophosphate, a substrate in the mitochondria of insect flight muscles, is not (Kobayashi, 1965). Yet, multiple forms of α-glycerophosphate dehydrogenase can be detected on gel electrophoresis of homogenates (Allen, 1968). The enzyme β-hydroxybutyrate dehydrogenase can be solubilized from the mitochondria. It requires only nicotinamide adenine dinucleotide (NAD^+), which is not replaceable by nicotinamide adenine dinucleotide phosphate ($NADP^+$) (Conger and Eichel, 1965). There are no significant changes in β-hydroxybutyrate dehydrogenase activity during the cycle of induced growth oscillation (Nishi and Scherbaum, 1962a).

Reduced NAD (NADH) oxidase is present in particles, presumably mitochondria, of *T. pyriformis* (Eichel, 1956a,b; Wingo and Thompson, 1960; Nishi and Scherbaum, 1962b; Rahat *et al.*, 1964); and multiple forms of the oxidase can be detected on gel electrophoresis of cell homogenates (Allen, 1968). This system transfers electrons from NADH via an electron transport mechanism to O_2. It is inhibited by high levels of antimycin A, a naphthoquinone (SN5949), α,α′-dipyridyl, cyanide, hydroxylamine, salicylaldoxamine, and diethyldithiocarbamate (Eichel, 1956a, b), as well as by amytal, chlorpromazine, and sulfide (Rahat *et al.*, 1964). Although antimycin A, which inhibits electron transport in mammalian cells, blocks NADH oxidase activity, intact mitochondria of *T. pyriformis* are resistant to this agent (Kobayashi, 1965), and intact cells are actually stimulated to divide in its presence (Shug *et al.*, 1968; Elson *et al.*, 1970). For intact cells or mitochondria, NADH oxidase activity is dependent on the presence of fumarate, malate, or oxalacetate. This dependency is due to the transport of hydrogen of NADH across the mitochondrial membrane by a malate–oxalacetate shuttle (Rahat *et al.*, 1964). The activity of NADH oxidase decreases following heat treatment to induce synchrony (Nishi and Scherbaum, 1962b); and, in homogenates, the system is very sensitive to heat (Eichel, 1956b), leading to speculation that it could be the temperature-sensitive controlling mechanism for synchrony.

Mammalian cytochrome c can be used to detect enzymes described as NADH–cytochrome c reductase (Eichel, 1956a,b; 1959a; Wingo and Thompson, 1960; Kamiya and Takahashi, 1961; Brightwell *et al.*, 1968) and reduced NADP (NADPH)–cytochrome c reductase (Brightwell *et al.*, 1968) in crude cell extracts of *T. pyriformis*. The NADH–cytochrome c reductase is inactivated by heat and by nicotinic acid, an agent which also prevents

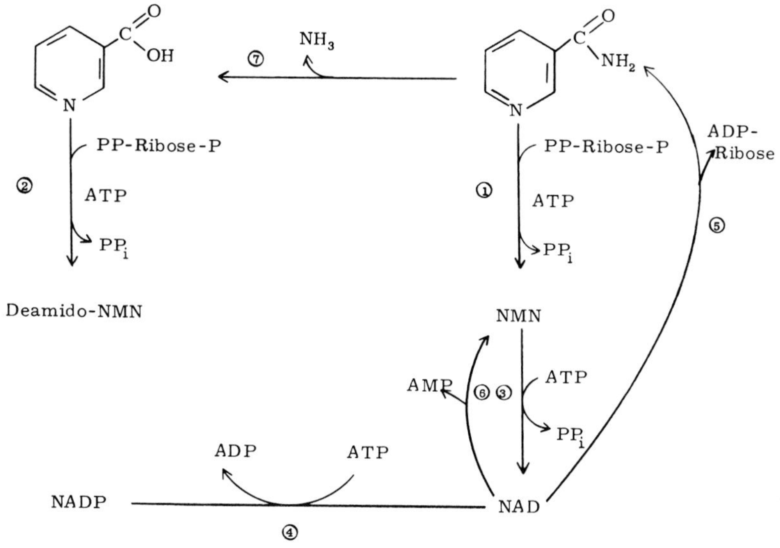

FIG. 4.3. Pyridine nucleotide metabolism of *Tetrahymena pyriformis*. ADP, adenosine diphosphate; ATP, adenosine triphosphate; AMP, adenosine monophosphate; NAD, nicotinamide adenine dinucleotide; NADP, nicotinamide adenine dinucleotide phosphate; NMN, nicotinamide mononucleotide; PP-ribose-P, 5-phosphoribosyl-1-pyrophosphate; PP_i, inorganic pyrophosphate.

synchronized cell division (Kamiya and Takahashi, 1961). The significance of these reductases is not understood, since very little cytochrome c oxidase can be detected, since O_2 serves as a better acceptor of electrons, and since a mammalian-type cytochrome c is not a component of the NADH oxidase system (Eichel, 1956a,c; 1959b).

A survey of pyridine nucleotide metabolism by *T. pyriformis* reveals that the enzymatic pathways are similar to those of higher animals (Fig. 4.3, reactions 1–7) (Belinsky and Dietrich, 1969). The enzymes present are nicotinamide mononucleotide (NMN) pyrophosphorylase (reaction 1), nicotinate mononucleotide pyrophosphorylase (reaction 2), NAD pyrophosphorylase (reaction 3), NAD kinase (reaction 4), NAD glycohydrolase (reaction 5), nucleotide pyrophosphatase (reaction 6), and nicotinamide deamidase (reaction 7). The presence of the pyrophosphatase had been noted previously (Villela, 1967). Table 4.1 includes the activities and the apparent Michaelis constants for these enzymes.

The levels of NADH and NADPH in exponentially growing cells are 0.36 and 0.46 nmoles/mg of protein, respectively (Nishi and Scherbaum, 1962b). A value for NAD^+ is 0.35 nmoles/gm of cells (Belinsky and Dietrich, 1969).

TABLE 4.1

ACTIVITY OF PYRIDINE NUCLEOTIDE-METABOLIZING ENZYMES IN *Tetrahymena pyriformis*[a,b]

Enzyme	Activity[c]	Apparent K_m
NMN pyrophosphorylase	0.013	5.0×10^{-4} *M* Nicotinamide
		5.0×10^{-5} *M* PP-Ribose-P
		2.0×10^{-3} *M* ATP
Nicotinate mononucleotide pyrophosphorylase	0.010	7.0×10^{-5} *M* Nicotinate
NAD pyrophosphorylase	0.579	2.0×10^{-3} *M* NMN
		1.0×10^{-3} *M* ATP
NAD kinase	0.780	7.0×10^{-3} *M* NAD
		3.0×10^{-3} *M* ATP
NAD glycohydrolase	330	3.0×10^{-4} *M* NAD
Nucleotide pyrophosphatase	54.9	
Nicotinamide deamidase	6.15	8.0×10^{-4} *M* Nicotinamide

[a] Data from Belinsky and Dietrich (1969).

[b] Abbreviations: NMN, nicotinamide mononucleotide; NAD, nicotinamide adenine dinucleotide; ATP, adenosine triphosphate; PP-ribose-P, 5-phosphoribosyl-1-pyrophosphate.

[c] Micromoles converted per hour per gram of cells.

A succinate oxidase system is found in *T. pyriformis* (Eichel, 1954, 1956c, 1959a; Edwin and Green, 1960). There is good evidence for the involvement of flavin in this process, but it is not required when the artificial electron acceptor, phenazine methosulfate, is used (Eichel, 1956c). Malonate inhibits respiration, probably by blocking succinate dehydrogenase (McCashland and Kronschnabel, 1962). Mammalian cytochrome c accepts electrons from succinate in the presence of extracts of *T. pyriformis* (Eichel, 1954; Seaman, 1954); but, as in the cases of NADH- and NADPH-cytochrome c reductase, the significance of this reaction is not known. Succinate has the property of stabilizing the succinate oxidase system and also NADH oxidase, NADH–cytochrome c oxidase, glutamate oxidase, β-hydroxybutyrate oxidase, and fumarase (Eichel, 1959a).

Cytochromes

Results regarding the cytochrome components of the electron transport system of *Tetrahymena pyriformis* have been confusing. An early report (Baker and Baumberger, 1941), from studies involving spectroscopic methods, states that cytochromes a, a_2, b, and c are present. Seaman (1949) also found cytochrome c. Ryley (1952) reported only weak absorption bands for cytochromes a, b, and c but a strong band at 552 nm described as cytochrome e, which was also observed by Møller and Prescott (1955). Kobayashi (1965) found cytochromes b and c, but not cytochrome a. Since cytochrome e absorption is very similar to that for cytochrome c, it is probable that the observers were looking at the same cytochrome. An oxidase for mammalian cytochrome c is either absent (Ryley, 1952) or extremely low (Eichel, 1959b), but Seaman (1954) insists that this activity can be found. He states that the oxidase is lost or inactivated during some procedures for preparing extracts, but Eichel (1956a) could not repeat Seaman's preparatory procedure.

The cytochromes absorbing at 553 nm (described as cytochrome c but probably identical to cytochrome e) and at 560 nm (cytochrome b) have been highly purified and some of their properties investigated (Yamanaka *et al.*, 1968). The first has an absorption maximum at 410 nm in the oxidized form and maxima at 414, 523, and 553 nm in the reduced form. The protein has an affinity for diethylaminoethyl (DEAE)–cellulose columns and is not precipitable with saturated ammonium sulfate. It is similar to "bacterial-type" cytochrome c and to *Euglena* cytochrome c-552 (Perini *et al.*, 1964) in that it reacts with *Pseudomonas* cytochrome oxidase but not with mammalian cytochrome oxidase. The second cytochrome shows a peak at 411 nm in the oxidized form and peaks at 424, 529, and 560 nm in the reduced form. It combines with CO and cyanide, but not with azide; and it has no peroxidase or catalase activity.

Cytochrome a, at the terminal portion of the respiratory chain, and cytochrome oxidase, which catalyzes the direct reduction of oxygen, cannot be detected in *T. pyriformis* by normal assay procedures (Ryley, 1952; Eichel, 1959a; Kobayashi, 1965). Cyanide and carbon monoxide, inhibitors of cytochrome oxidase, are reported to have only a small effect on the respiration of this organism (Ryley, 1952); and cells can be adapted to grow in the presence of relatively high amounts of cyanide (McCashland, 1955, 1956, 1963; McCashland and Steinacher, 1962; McCashland *et al.*, 1957; van de Vijver, 1966b). (The resistance to cyanide is not a permanent genetic change, for cells lose this property when grown in the absence of the agent.) All of this evidence makes a potent argument for the absence of cytochrome a and cytochrome oxidase. However, cyanide does inhibit, in isolated mitochondria, the NADH oxidase system (Eichel, 1956b; Rahat *et al.*, 1964) as well as succinate, glutamate, and ascorbate oxidase (Kobayashi, 1965). Azide, another inhibitor of cytochrome oxidase, inhibits respiration (Ryley, 1952) and also blocks synchronous division (Hamburger, 1962). Recently, an ascorbate–N,N,N',N'-tetramethyl-p-phenylenediamine system which uses oxygen as the terminal acceptor of electrons has been found (Perlish and Eichel, 1968, 1969). This system is sensitive to cyanide, to azide, and to other inhibitors of succinate and NADH oxidase. Also, recently reported is a cytochrome component, presumably cytochrome a, which uses oxygen more rapidly than cytochromes b and c (Turner *et al.*, 1969). In view of these reports, it now appears that a cytochrome oxidase system, however unusual, is present in *T. pyriformis*.

There is no evidence that β-hydroxybutyrate, glutamate, and the Krebs cycle intermediates are oxidized by different electron transport systems. Very likely, succinate and ascorbate share a portion of the pathway. A possible electron transport system for *Tetrahymena* is as follows.

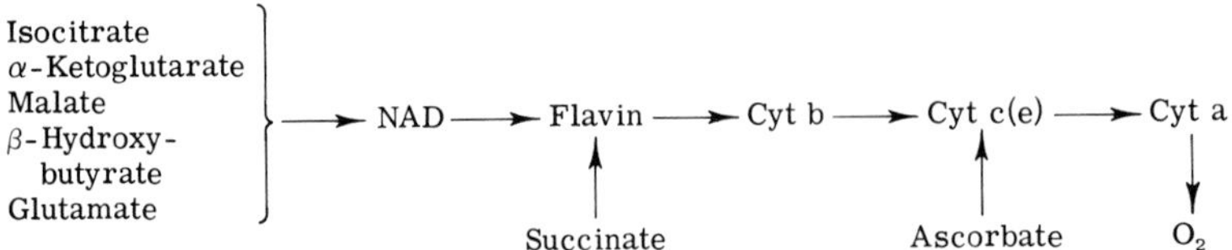

In addition to absorption bands for cytochromes, bands for hemoglobin can also be detected in *T. pyriformis* (Ryley, 1952; Keilin and Ryley, 1953, 1957); but the function of hemoglobin in the organism remains obscure. Ubiquinones (coenzymes Q) are also present (Taketomi, 1961), but it is not certain that they participate in the electron transport sequence.

Uncoupling Agents

2,4-Dinitrophenol is an inhibitor that uncouples the oxidative phosphorylation of adenosine diphosphate (ADP) to ATP and allows respiration to proceed at an increased rate. Such an effect by this compound is seen on *Tetrahymena pyriformis* (Ryley, 1952; Hamburger and Zeuthen, 1957; Kobayashi, 1965). 2,4-Dinitrophenol also inhibits the growth of this organism (Conner, 1957) and delays the division of synchronized cells (Rasmussen and Zeuthen, 1962; Hamburger and Zeuthen, 1957). The inhibition of growth is alleviated by stigmasterol, but the reversing effect is influenced by the presence in the medium of phosphate, Tween 80, acetate, amino acids, purines, and pyrimidines (Conner, 1958). The effect of phosphate appears to be related to the fact that 2,4-dinitrophenol inhibits phosphate uptake by the cell, a process reversed by stigmasterol (Conner, 1957, 1959; Conner *et al.*, 1961). The presence of glucose or cholesterol in the medium prevents loss of phosphate from the cells but does not affect uncoupling by the agent (Conner and Longobardi, 1963).

Chlorpromazine, pentachlorophenol, and tributyl tin also act as uncoupling agents (Kobayashi, 1965); and oxidative phosphorylation is decreased by the heat treatment used to induce synchronous division (Nishi and Scherbaum, 1962a; Rooney and Eiler, 1969).

Studies of electron transport in *T. pyriformis* are hampered by an agent which inhibits this process and also uncouples oxidative phosphosphorylation. Eichel (1959b, 1960) believes that it is a fatty acid or a lysophosphatide, both of which are produced by phospholipase A. Its effect is prevented by bovine serum albumin (Eichel, 1960), which is known to complex with fatty acids, and by various tocopherols (Edwin and Green, 1960). This inhibitory lipid may be responsible for the low P : O ratios found for oxidative phosphorylation by mitochondria from *T. pyriformis* (Eichel and Rem, 1961; Nishi and Scherbaum, 1962a). However, mitochondria carefully prepared in the presence of bovine serum albumin show P : O ratios greater than 1 for succinate, α-ketoglutarate, fumarate, malate, glutamate, citrate, β-hydroxybutyrate, pyruvate, and lactate (Kobayashi, 1965).

Utilization of Energy

Much of the ATP produced by *Tetrahymena pyriformis* is used for motility through the beating of the cilia. Evidence is available to show that a certain intracellular level of ATP is required for the maintenance of motility (Burnasheva and Efremenko, 1962; Burnasheva and Karusheva, 1966). Some inhibitors of motility are physostigmine and diisopropylfluorophosphate (Seaman, 1951), as well as hexamethonium and an antibody prepared from isolated cilia (Warnock, 1962; Warnock and van Eys, 1963; van Eys and Warnock, 1963). Hexamethonium leads to a small increase in the intracellular ATP content. Its effect can be reversed by acetylcholine, which was at one time thought to be involved in impulse transmission along the rows of cilia (Seaman, 1951). The fact that ATP produced by respiration and that produced by glycolysis can be used for motility indicates that separate compartments for glycolytic ATP and respiratory ATP do not exist (Warnock, 1962).

Peroxisomes

Mitochondria are not the only particles accomplishing oxidation in *Tetrahymena pyriformis*. Other microbodies, now called "peroxisomes," contain enzymes catalyzing the oxidation of α-hydroxy and α-D-amino acids (Eichel and Rem, 1959a,b; Baudhuin *et al.*, 1965; Levy and Wasmuth, 1970). The α-hydroxy acid oxidase is distinct from the particle-bound, NAD^+-linked lactate dehydrogenase; the two can be separated by treatment with 0.1% cetyldimethylbenzylammonium chloride (Eichel *et al.*, 1964; Eichel, 1966). With lactate as substrate, the oxidase reaction proceeds as follows:

$$\text{Lactate} + O_2 \longrightarrow \text{Pyruvate} + H_2O_2$$

The system reduces 2,6-dichlorophenolindophenol, but reacts poorly with ferricyanide and is not inhibited by cyanide or other inhibitors of the NADH and succinate oxidase systems (Eichel *et al.*, 1964; Eichel and Rem, 1962). The hydrogen peroxide formed by the enzyme is undoubtedly destroyed by the catalase present in this ciliate (Roth and Eichel, 1955; Roth and Buccino, 1965; Chou *et al.*, 1970). Peroxisomes of *T. pyriformis* can be visualized by electron microscopy (Williams and Luft, 1968; Müller, 1969).

REFERENCES

Allen, S. L. (1968). *Ann. N.Y. Acad. Sci.* **151,** 190.
Baker, E. G. S., and Baumberger, J. P. (1941). *J. Cell. Comp. Physiol.* **17,** 285.
Baudhuin, P., Müller, M., Poole, B., and de Duve, C. (1965). *Biochem. Biophys. Res. Commun.* **20,** 53.
Belinsky, C., and Dietrich, L. S. (1969). *Biochim. Biophys. Acta* **177,** 668.
Bhagavan, N. V., and Eiler, J. J. (1961). *Fed. Proc. Fed. Amer. Soc. Exp. Biol.* **20,** 170.
Brightwell, R., Lloyd, D., Turner, G., and Venables, S. E. (1968). *Biochem. J.* **109,** 42P.
Burnasheva, S. A., and Efremenko, M. V. (1962). *Biokhimiya* **27,** 167.
Burnasheva, S. A., and Karusheva, T. P. (1966). *Biokhimiya* **32,** 222.

Byfield, J. E., Chou, S. C., and Scherbaum, O. H. (1962). *Biochem. Biophys. Res. Commun.* **9,** 226.
Chaix, P., Chauvet, J., and Fromageot, C. (1947). *Antonie van Leeuwenhoek J. Microbiol. Serol.* **12,** 145.
Chou, S. C., Yamada, K., Conklin, K. A., and Hokama, Y. (1970). *Fed. Proc. Fed. Amer. Soc. Exp. Biol.* **29,** 493.
Conger, N. E., and Eichel, H. J. (1965). *Fed. Proc. Fed. Amer. Soc. Exp. Biol.* **24,** 350.
Conner, R. L. (1957). *Science* **126,** 698.
Conner, R. L. (1958). *J. Protozool.* **5,** Suppl. 25.
Conner, R. L. (1959). *J. Protozool.* **6,** Suppl. 10.
Conner, R. L., and Cline, S. G. (1967). *J. Protozool.* **14,** 22.
Conner, R. L., and Longobardi, A. E. (1963). *J. Protozool.* **10,** Suppl. 8.
Conner, R. L., Kornacker, M. S., and Goldberg, R. (1961). *J. Gen. Microbiol.* **26,** 437.
Danforth, W. F. (1967). *In* "Research in Protozoology" (T-T. Chen, ed.), Vol. 1, p. 201. Pergamon Press, Oxford.
Edwin, E. E., and Green, J. (1960). *Arch. Biochem. Biophys.* **87,** 337.
Eichel, H. J. (1954). *J. Biol. Chem.* **206,** 159.
Eichel, H. J. (1956a). *J. Biol. Chem.* **222,** 121.
Eichel, H. J. (1956b). *J. Biol. Chem.* **222,** 137.
Eichel, H. J. (1956c). *Biochim. Biophys. Acta* **22,** 571.
Eichel, H. J. (1959a). *Biochem. J.* **71,** 106.
Eichel, H. J. (1959b). *Biochim. Biophys. Acta* **34,** 589.
Eichel, H. J. (1960). *Biochim. Biophys. Acta* **43,** 364.
Eichel, H. J. (1966). *Biochim. Biophys. Acta* **128,** 183.
Eichel, H. J., and Rem, L. T. (1959a). *Biochim. Biophys. Acta* **35,** 571.
Eichel, H. J., and Rem, L. T. (1959b). *Arch. Biochem. Biophys.* **82,** 484.
Eichel, H. J., and Rem, L. T. (1961). *Progr. Protozool. Proc. Int. Congr. Protozool.* **1,** 148.
Eichel, H. J., and Rem, L. T. (1962). *J. Biol. Chem.* **237,** 940.
Eichel, H. J., Goldenberg, E. K., and Rem, L. T. (1964). *Biochim. Biophys. Acta* **81,** 172.
Eiler, J. J., Krezanoski, J. Z., and Lee, K. H. (1959). *J. Amer. Pharm. Ass. Sci. Ed.* **48,** 290.
Elliott, A. M., and Bak, I. J. (1964). *J. Cell. Biol.* **20,** 113.
Elliott, A. M., Travis, D. M., and Bak, I. J. (1962). *Biol. Bull.* **123,** 487.
Elliott, A. M., Travis, D. M., and Work, J. A. (1966). *J. Exp. Zool.* **161,** 177.
Elson, C., Hartman, H. A., Shug, A. L., and Shrago, E. (1970). *Science* **168,** 385.
Hamburger, K. (1962). *C. R. Trav. Lab. Carlsberg* **32,** 359.
Hamburger, K., and Zeuthen, E. (1957). *Exp. Cell Res.* **13,** 443.
Kamiya, T., and Takahashi, T. (1961). *J. Biochem. (Tokyo)* **50,** 277.
Keilen, D., and Ryley, J. F. (1953). *Nature (London)* **172,** 451.
Keilen, D., and Ryley, J. F. (1957). *Nature (London)* **179,** 988.
Kobayashi, S. (1965). *J. Biochem. (Tokyo)* **58,** 444.
Levy, M. R., and Elliott, A. M. (1968). *J. Protozool.* **15,** 208.
Levy, M. R., and Wasmuth, J. J. (1970). *Biochim. Biophys. Acta* **201,** 205.
Lwoff, M. (1938). *C. R. Acad. Sci. Ser.* **D.127,** 1170.
McCashland, B. W. (1955). *J. Protozool.* **2,** 97.
McCashland, B. W. (1956). *J. Protozool.* **3,** 131.
McCashland, B. W. (1963). *Growth* **27,** 47.
McCashland, B. W., and Kronschnabel, J. M. (1962). *J. Protozool.* **9,** 276.
McCashland, B. W., and Steinacher, R. H. (1962). *Proc. Soc. Exp. Biol. Med.* **111,** 789.

McCashland, B. W., Marsh, W. R., and Kronschnabel, J. M. (1957). *Growth* **21,** 21.
Møller, K. M., and Prescott, D. M. (1955). *Exp. Cell Res.* **9,** 373.
Müller, M. (1969). *Ann. N.Y. Acad. Sci.* **168,** 292.
Müller, M., Hogg, J. F., and de Duve, C. (1968). *J. Biol. Chem.* **243,** 5385.
Nishi, A., and Scherbaum, O. H. (1962a). *Biochim. Biophys. Acta* **65,** 419.
Nishi, A., and Scherbaum, O. H. (1962b). *Biochim. Biophys. Acta* **65,** 411.
Ormsbee, R. A. (1942). *Biol. Bull.* **82,** 423.
Pace, D. M., and Ireland, R. L. (1945). *J. Gen. Physiol.* **28,** 547.
Pace, D. M., and Lyman, E. D. (1947). *Biol. Bull.* **92,** 210.
Parsons, J. A. (1964). *J. Cell Biol.* **23,** 70A.
Perini, F., Kamen, M. D., and Schiff, J. A. (1964). *Biochim. Biophys. Acta* **88,** 74.
Perlish, J. S., and Eichel, H. J. (1968). *J. Protozool.* **15,** Suppl. 15.
Perlish, J. S., and Eichel, H. J. (1969). *J. Protozool.* **16,** Suppl. 12.
Rahat, M., Judd, J., and van Eys, J. (1964). *J. Biol. Chem.* **239,** 3537.
Rasmussen, L. (1963). *C. R. Trav. Lab. Carlsberg* **33,** 53.
Rasmussen, L., and Zeuthen, E. (1962). *C. R. Trav. Lab. Carlsberg* **32,** 333.
Rooney, D. W., and Eiler, J. J. (1969). *J. Cell Biol.* **41,** 145.
Roth, J. S. (1962). *J. Protozool.* **9,** 142.
Roth, J. S., and Buccino, G. (1965). *J. Protozool.* **12,** 432.
Roth, J. S., and Eichel, H. J. (1955). *Biol. Bull.* **108,** 308.
Roth, J. S., and Eichel, H. J. (1961). *J. Protozool.* **8,** 69.
Roth, J. S., Eichel, H. J., and Ginter, E. (1954). *Arch. Biochem. Biophys.* **48,** 112.
Ryley, J. F. (1952). *Biochem. J.* **52,** 483.
Sato, H. (1960). *Anat. Rec.* **138,** 381.
Scherbaum, O. H., Chou, S. C., Seraydarian, K. H., and Byfield, J. E. (1962). *Can. J. Microbiol.* **8,** 753.
Seaman, G. R. (1949). *J. Cell. Comp. Physiol.* **33,** 441.
Seaman, G. R. (1951). *J. Cell. Comp. Physiol.* **37,** 309.
Seaman, G. R. (1954). *Arch. Biochem. Biophys.* **48,** 424.
Sedar, A. W., and Rudzinska, M. A. (1956). *J. Biophys. Biochem. Cytol.* **2,** Suppl. 331.
Shug, A. L., Ferguson, S., and Shrago, E. (1968). *Biochem. Biophys. Res. Commun.* **32,** 81.
Taketomi, T. (1961). *Z. Allg. Mikrobiol.* **1,** 331.
Turner, G., Lloyd, D., and Chance, B. (1969). *Biochem. J.* **114,** 91P.
van de Vijver, G. (1966a). *Enzymologia* **31,** 363.
van de Vijver, G. (1966b). *Enzymologia* **31,** 382.
van de Vijver, G. (1967). *Enzymologia* **33,** 331.
van Eys, J., and Warnock, L. G. (1963). *J. Protozool.* **10,** 465.
Villela, G. G. (1967). *Nature (London)* **213,** 79.
Warnock, L. G. (1962). *Diss. Abstr.* **23,** 3623.
Warnock, L. G., and van Eys, J. (1963). *J. Cell Comp. Physiol.* **61,** 309.
Williams, N. E., and Luft, J. H. (1968). *J. Ultrastruct. Res.* **25,** 271.
Wingo, W. J., and Thompson, W. D. (1960). *Fed. Proc. Fed. Amer. Soc. Exp. Biol.* **19,** 243.
Yamanaka, T., Nagata, Y., and Okunuki, K. (1968). *J. Biochem. (Tokyo)* **63,** 753.

CHAPTER 5

Amino Acid and Protein Metabolism

Introduction

Proteins, which consist of long chains of the 20 naturally occurring amino acids, have extreme diversity, allowing each animal species and each tissue to have its own specific kinds. The architecture and functioning of living matter, is, to a large extent, dependent on the proteins present. More than half the dry tissue of animals is protein; and these molecules are the structural components of hair, wool, collagen, muscle fibers, and cilia. Enzymes are also proteins, and they catalyze nearly all of the chemical reactions of cells. Protein synthesis, in all organisms, involves a series of complex, enzymatic reactions, about which a great deal is known for bacteria and mammalian cells but very little for *Tetrahymena*.

Some of the amino acids comprising cellular protein are required in the diet of animals, for animals have lost the ability to make the carbon skeletons. The amino acid requirements for *Tetrahymena* have been extensively investigated; but except for one case, the enzymes involved in its metabolism of amino acids have received scant attention. Amino acids can be precursors of other important nitrogenous compounds, such as porphyrins and catecholamines, as well as the catabolic products ammonia, urea, allantoin, uric acid, and creatinine. *Tetrahymena* use amino acids for various synthetic purposes and also degrade them to ammonia, the major excretory form of nitrogen for this organism.

Amino Acid Requirements

In regard to amino acids, *Tetrahymena pyriformis* has an animal-like nutritional requirement. Those needed for growth are methionine, arginine, threonine, tryptophan, valine, isoleucine, leucine, phenylalanine, histidine, and lysine (Kidder and Dewey, 1951; Dewey and Kidder, 1960a). The same requirement is shared by the ciliates *Paramecium* and *Glaucoma* (Holz, 1969) and by the growing rat. Dewey and Kidder (1960a), on examination of a large number of strains of *T. pyriformis*, find only one that differs in that it requires an exogenous supply of serine; but other investigators (Elliott and Hayes, 1955; Elliott and Clark, 1958a,b) find that several wild strains require this amino acid. However, it is reported that one such clone produces serine endogenously—a situation that is not easily explained (Elliott, 1959). Some other newly isolated strains are reported to require alanine, glycine, proline, aspartate, and glutamate in addition to the usual 10 amino acids (Elliott *et al.*, 1962, 1964); but after a period of time in culture, some of them are said to retain only the normal requirements. With the present lack of definitive data, it is doubtful if any of these strains of *T. pyriformis* had a true requirement for amino acids other than those commonly needed.

Most other species of *Tetrahymena* which have been studied require only the 10 amino acids listed above and, perhaps, serine. The strains are *T. setifera* (Holz *et al.*, 1962), *T. vorax* (Kessler, 1961; Shaw and Williams, 1963), *T. corlissi* (Holz, 1964), *T. patula* (Holz, 1964), and *T. paravorax* (Holz *et al.*, 1961). A glycine requirement is reported for *T. patula* (Kessler, 1961) and for *T. paravorax* (Holz *et al.*, 1961), but in both cases the requirement could likely be removed by increasing the folic acid concentration in the medium. Such an addition would presumably increase the activity of serine hydroxymethyltransferase, which produces glycine from serine (Dewey and Kidder, 1960a).

The common amino acids which are not required for growth are stimulatory in a nonspecific fashion, providing a source of carbon and nitrogen for the organism and sparing the requirement of some essential amino acids.

Proteinases

Growing cultures of *Tetrahymena pyriformis* easily degrade proteins. The cells are able to liquefy gelatin (Lwoff, 1932), and extracts of the organism digest casein, egg albumin, and α-glutelin (Lawrie, 1937) as a result of proteinase activity.

Two extracellular and one intracellular proteinases have been purified and studied (Viswanatha and Liener, 1955, 1956; Dickie and Liener, 1962a,b). One of the extracellular enzymes is depressed in the presence of 1% glucose, but it can be isolated as the crystalline mercuric derivative. It is activated by cysteine or ethylenediaminetetraacetate (EDTA) and splits denatured proteins or the synthetic substrates, carbobenzoxy-L-glutamyl-L-tyrosine and α-benzoyl-L-arginine methyl ester. Tyrosine and, probably, tryptophan are absent from the enzyme molecule (Viswanatha and Liener, 1956). The intracellular proteinase is more active than the extracellular ones, but all three have a low specificity for the various amide bonds of insulin (Dickie and Liener, 1962b) and have histidine as the N-terminal amino acid (Dickie and Liener, 1962a).

It has been reported that proteinase activity decreases in *T. pyriformis* under starvation conditions (Lantos *et al.*, 1964), but this has been disputed by Levy and Elliott (1968), who find an increase in this activity in the absence of nutrients.

Since *T. pyriformis* has active proteinases and requires the same amino acids as mammals, the organism has been put to the practical use of evaluating protein quality (Dunn and Rockland, 1947; Rockland and Dunn, 1949; Anderson and Williams, 1951; Pilcher and Williams, 1954; Fernell and Rosen, 1956; Rosen and Fernell, 1956; Celliers, 1961; Teunisson, 1961; Stott *et al.*, 1963; Stott and Smith, 1966; Boyne *et al.*, 1967; Kamath and Ambegakar, 1968; Bergner *et al.*, 1968). In general, the organism reacts like the rat. Vegetable proteins are inferior to animal proteins, and overheating adversely affects the quality.

The amino acids produced by protein breakdown are subject to degradation for a source of energy. Adaptive formation of amino acid decarboxylases has been reported (Mefferd *et al.*, 1952), and Seaman (1955), in a review article, presents two tables showing that all the common amino acids can be transaminated. Some amino acids stimulate respiration (Roth *et al.*, 1954). The most effective is L-phenylalanine, followed by L-tyrosine, L-cysteine, D-cysteine, L-proline, L-isoleucine, L-leucine, and D-leucine. The remaining naturally occurring amino acids and L-homocysteine and α-aminobutyrate are not stimulatory. L-Amino acid oxidase activity cannot be demonstrated in cell homogenates (Roth *et al.*, 1954), but D-amino acid oxidase activity is present in subcellular particles known as peroxisomes (Baudhuin *et al.*, 1965; Müller *et al.*, 1968.

Protein Biosynthesis

Amino acid activation and incorporation into protein by the microsomal system of *Tetrahymena pyriformis* apparently proceeds in a manner similar to that in animals. An early investigation showed that a microsomal preparation with added soluble enzymes could incorporate amino acids into a linkage which is resistant to hot trichloroacetic acid and to ribonuclease (Mager and Lipmann, 1958); but, except for the demonstration of enzymes activating L-valine (Rosenbaum and Holz, 1966; Plesner, 1961) and L-leucine (Suyama and Eyer, 1967), very little has been done to characterize the enzymes of *T. pyriformis* involved in protein synthesis. As expected, the microsomal system for incorporation of phenylalanine is dependent on an energy-generating system and on polyuridylic acid (Letts and Zimmerman, 1970).

An interesting anomaly in the ribosomal protein synthesis system in *T. pyriformis* is that the monoribosomes may exist in two different states depending on the conditions of growth. Monoribosomes from normal cells have a sedimentation coefficient of 80 to 82 S (Whitson *et al.*, 1966; Leick and Plesner, 1968a; Chi and Suyama, 1970), but after heat treatment of cells to produce synchronized division, the value is 70 S (Plesner, 1961; Scherbaum, 1963; Leick and Plesner, 1968a). In comparison, the sedimentation value of bacterial ribosomes is 70 S and that for other eukaryotes is 80 S (Leon and Mahler, 1968). The 82 S ribosomes of *T. pyriformis* have a molecular weight of 4.6×10^6 daltons, and are composed of ribonucleic acid and protein in a ratio of 1:1 (Weller *et al.*, 1968). Their density is 1.56 gm/cm^3 after formaldehyde fixation (Chi and Suyama, 1970). The monoribosomes consist of one 50 S subunit containing 25 and 5 S ribonucleic acid and one 30 S subunit with 17 S ribonucleic acid (Leick and Plesner, 1968a). Thirteen or more different protein bands can be distinguished on gel electrophoresis. The spermidine and putrescine present in *T. pyriformis* (Arlock *et al.*, 1969) are bound, to a great extent, to the intact ribosome (Weller *et al.*, 1968).

Although most ribosomes are not attached to the endoplasmic reticulum of *T. pyriformis* (Cameron *et al.*, 1966), they can exist as larger complexes (polyribosomes) (Fig. 5.1). Normal cells contain ribosomes with sedimentation coefficients of 75 (82?), 105, 130, and 170 S (Kumar, 1968). Cameron *et al.* (1966) found ribosomes with sedimentation coefficients of 72 (82?), 117, 148, 186, and 203 S; and Scherbaum (1963) found ribosomes with a value of 125 S. After release from heat shock, an accumulation of heavy

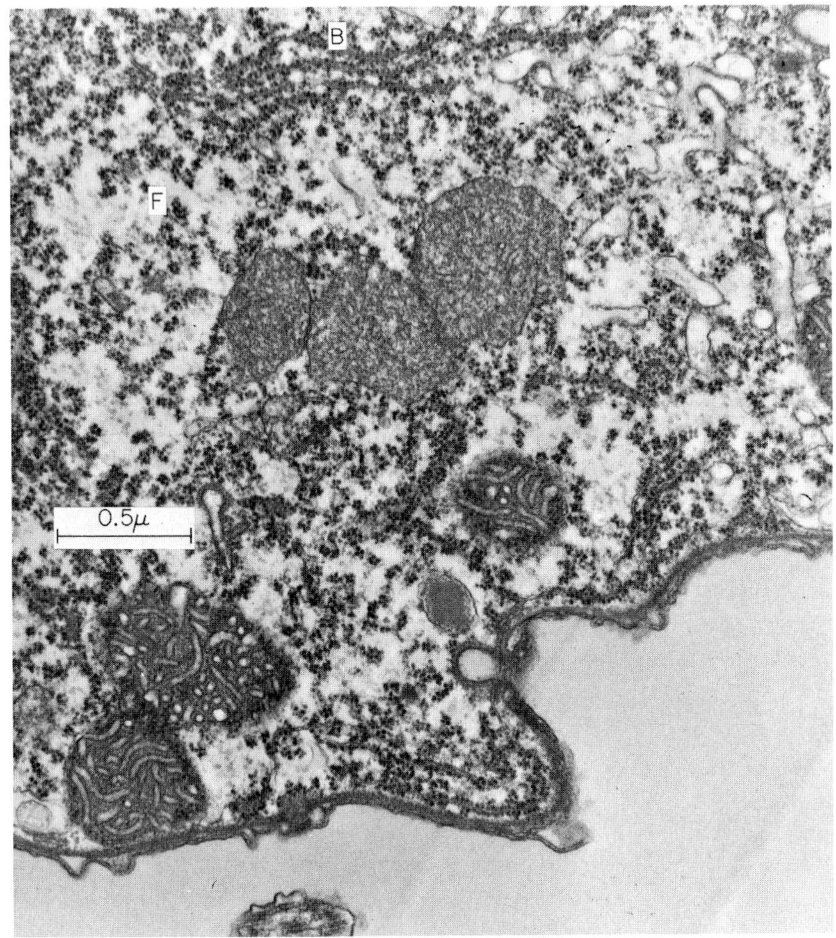

FIG. 5.1. Polysomes, free (F) and membrane-bound (B). Magnification: ×55,000. Mitochondria are in the early stationary phase. (Courtesy of Dr. I. L. Cameron and Mr. Glenn Williams.)

polyribosomes is evident in the sucrose-gradient profile (Hartmann and Dowben, 1970b; Hermolin and Zimmerman, 1969).

Although Leick and Plesner (1968a,b) use freezing followed by thawing, homogenization, and centrifugation to prepare ribosomes, Lyttleton (1963) recommends rupture of cells by indole-saturated water. A still better procedure involves breaking the cells by nitrogen cavitation in a medium containing calcium ions (Hartmann and Dowben, 1970a).

Studies have been made on the formation of ribosomes in *T. pyriformis* (Leick and Plesner, 1968a,b), a process occurring in the nucleoli (Leick, 1969b). The 50 S subunits show a delayed kinetics of labeling with ^{14}C-uridine, a precursor of ribonucleic acid, as compared to the 30 S subunits. This observation is in agreement with the findings in mammalian systems (see Perry, 1967). The delay in labeling of the 50 S subunit is a result of a slower rate at some stage later than the transcription of 25 S ribonucleic acid, since 25 and 17 S ribonucleic acids are formed at equal rates (Leick and Plesner, 1968a; Leick, 1969b). The delay presumably involves delayed completion as well as delayed transport of large portions of the 50 S subunit.

Rapidly labeled ribonucleoprotein particles with sedimentation values of 40 (45) and 60 S can be isolated from extracts of exponentially growing *T. pyriformis* incubated with ^{14}C-uridine (Leick and Plesner, 1968b; Kumar, 1968, 1970). The 40 S particles exist solely in the free form, whereas 60 S particles exist free as well as bound to larger entities sedimenting in the polysome region in a sucrose gradient. The bound 60 S ribonucleoprotein particles can be liberated by lowering the magnesium ion concentration during analysis. The 60 and 40 S subunits have densities of 1.57 and 1.53 gm/cm^3, respectively, after formaldehyde fixation (Chi and Suyama, 1970). These are the only subunits reported by Chi and Suyama (1970) for the 80 S ribosomes; they did not see the 50 and 30 S particles. Evidence that the 60 S particles contain predominantly 17 S ribonucleic acid and that the specific activity of the ribonucleic acid in the 60 and 40 S particles is higher than that in the 80 S particles (Leick and Plesner, 1968b) leads to the conclusion that the 40 S particles are precursors of 30 S subunits and that the 60 S particles are precursors of 50 S subunits. The nascent 40 and 60 S particles contain accessory proteins (about 15–20% of the total) which are split off when the particles are converted into their corresponding ribosomal subunits (Leick *et al.*, 1970). There are indications that the subribosomal particles do not participate in subunit exchange between each round of translation, as do the 30 and 50 S subunits of bacteria and yeast (Leick *et al.*, 1970).

A hybrid monosome made from the 60 S subunit from rat ribosomes and the 40 S subunit from *T. pyriformis* ribosomes is active in protein synthesis, but the 60 S subunit from *T. pyriformis* does not combine with any of a number of 40 S subunits from various species (Martin, 1968, 1969; Martin and Wool, 1969).

The microsomal system for protein synthesis is insensitive to chloramphenicol (Mager, 1960). Since this antibiotic arrests the growth of *T. pyriformis* and produces a 30–60% inhibition of amino acid incorporation into protein, another protein-synthesizing system must be sensitive to its action. Such a system, inhibited by chloramphenicol and by chlortetracycline but not by cycloheximide, is present in the mitochondria (Mager, 1960; Rosenbaum and Holz, 1966; Chi and Suyama, 1970). The activation of valine occurs in washed mitochondria, but not in washed cilia, pellicles, microsomes, or macronuclei (Rosenbaum and Holz, 1966). The mitochondrial incorporation of amino acids is dependent on adenosine triphosphate (ATP) and is about 20% as active (expressed per milligram of protein) as the microsomal system (Mager, 1960). Chi and Suyama (1970) concluded that some of the mitochondrial ribosomal proteins are made in the mitochondria. Suyama and Eyer (1967) found that the mitochondrial leucyl transfer ribonucleic acid and its activating enzyme are different from the corresponding extramitochondrial entities.

Other observations with chloramphenicol are that it delays synchronous division in heat-treated cells (Holz *et al.*, 1963) and that if such cells are transferred to inorganic medium with chloramphenicol, the agent has no effect on division (Lee *et al.*, 1959). Thus, nutrients are required for chloramphenicol to be effective against heat-treated cells. Growth of *T. pyriformis* in the presence of chloramphenicol leads to a prolonged lag phase and a longer mean generation time (Turner and Lloyd, 1970). Mitochondria become greatly decreased in size and more numerous, and they lose all of their oxidative capacity and respiratory control.

Mitochondrial ribosomes with a sedimentation coefficient of 80 S and indistinguishable from cytoplasmic ribosomes in sedimentation analysis are present in *T. pyriformis* (Chi and Suyama, 1970). These ribosomes differ, however, in that they produce only one (55 S) peak at low levels of magnesium ion. This peak can be separated into two subunits on CsCl density gradient centrifugation. There are also some electrophoretic differences on polyacrylamide gel columns between proteins isolated from mitochondrial and cytoplasmic 80 S ribosomes. The mitochondrial ribosomes contain relatively small 21 and 14 S ribonucleic acid molecules, which apparently function in protein synthesis. Calculated molecular weights for mitochondrial and cytoplasmic ribosomes are 3.24 to 3.26 $\times$ 10^6 daltons and 3.38 to 3.46 $\times$ 10^6 daltons, respectively (Chi and Suyama, 1970). Both types of ribosome are visible on electron microscopy (Swift *et al.*, 1964).

Nuclear ribonucleoprotein particles, containing predominantly 25 S ribonucleic acid, can be isolated from the nuclear envelope (Leick, 1969b). A major fraction of the particles sediments with a value of 60 S. These ribosomes, distinct from those of the cytoplasm, may be the site of yet unproved nuclear protein synthesis (Gorovsky, 1969).

Yet another protein-synthesizing system is reported by Seaman (1962).

This is in the kinetosomes, the basal bodies of cilia, which are thought by some to be self duplicating (Chatton and Lwoff, 1935; Allen, 1968a). The following requirements are reported: amino acids, inorganic phosphate, uridine monophosphate (UMP), cytidine monophosphate (CMP), guanosine monophosphate (GMP), thymine, inosine, $MgSO_4$, KF, hexose diphosphate, ATP, folic acid, riboflavin, vitamin B_{12}, biotin, pyridoxamine, and pantothenic acid. The system is reportedly not sensitive to chloramphenicol, but is sensitive to ribonuclease and deoxyribonuclease. If the hodgepodge mixture of chemical requirements of this system is not enough to cast enormous doubt on the validity of this observation, the battery of challenges to the purity of Seaman's (1960) preparation of kinetosomes should suffice (Rampton, 1962; Kimball, 1964; Pitelka and Child, 1964; Argetsinger, 1965; Hoffman, 1965; Satir and Rosenbaum, 1965).

Actinomycin D, an agent which has a primary effect on the transcription of ribonucleic acid (Hurwitz *et al.*, 1962) and a secondary effect on protein synthesis, binds to the deoxyribonucleic acid of *T. pyriformis* and other organisms (Müller and Crothers, 1968). By blocking messenger ribonucleic acid synthesis, it inhibits ribosome formation (Leick, 1969a); but there is an increase in the number of single 82 S ribosomes, and anomalous 70 S ribosomes also appear, apparently from the breakdown of polyribosomes (Whitson *et al.*, 1966). In heat-treated cultures of *T. pyriformis*, actinomycin D reduces total protein synthesis only slightly, which is an indication of its primary effect on ribonucleic acid synthesis (Byfield and Scherbaum, 1967a,b); but it can block division (Moner, 1964; Nachtwey and Dickinson, 1967) or the development of oral structures (Gavin and Frankel, 1969) if added at the appropriate time.

Cycloheximide is a strong, presumably specific, inhibitor of protein synthesis in *T. pyriformis*, as it is in other organisms. The agent rapidly and completely inhibits incorporation of amino acids into protein of synchronized cells, and oral development is stopped immediately if it is in the early stages (Gavin and Frankel, 1969; Frankel, 1969a, 1970b). Maximum delay in the time of cell division is observed if low levels of cycloheximide are added just prior to a "transition point," which is 55 min after the end of treatment to induce synchrony (Frankel, 1969a, 1970b). High levels completely inhibit division. Cells dividing synchronously, when treated with low levels of cycloheximide, continue to divide synchronously after a delay (Frankel, 1970a). Recovery from inhibition is not due to degradation of the drug—the cells either adapt to the presence of cycloheximide or exclude it.

Puromycin is a potent inhibitor of protein synthesis in *T. pyriformis* (Frankel, 1967). The growth of the organism is strongly inhibited by puromycin, but not by aminonucleoside or by methoxyphenylalanine (Bortle and Oleson, 1954). The inhibition of growth is relieved by peptides (Dewey and Kidder, 1964) which have the following structural requirements: (*1*) the natural form of the amino acid must be present as the N-terminal residue; (*2*) the N-terminal should preferably be glycine; and (*3*) the C-terminal residue should be relatively nonpolar. These studies suggest that puromycin interferes with the uptake of free amino acids but not that of peptides and that intracellular peptides are hydrolyzed by an aminopeptidase relatively specific for glycyl peptides. The free intracellular amino acids released by the peptidase are then able to exchange for free extracellular amino acids, a process presumably insensitive to puromycin. Recently, it has been noted that sterols and phosphatidyl choline are more potent than peptides in reversing the inhibition (Dewey, 1967). This leads one to believe that it is the transport of puromycin across the cell membrane that is blocked by peptides and lipids. Inside the cell, puromycin probably acts as an analog of transfer ribonucleic acid and inhibits protein synthesis by binding to the growing polypeptide chain, as it does in other cells (Nathans, 1964).

Some other observations on the effect of puromycin on *T. pyriformis* are that it delays synchronized division if added at the proper time (Cerroni and Zeuthen, 1962; Holz *et al.*, 1963; Frankel, 1967) and that it blocks regeneration of cilia (Child, 1965; Rosenbaum and Carlson, 1969). Also, development of oral primordia is arrested, and primordia in the early stages of membranelle differentiation are resorbed (Frankel, 1967, 1969b, 1970b). In this regard, one effect of puromycin appears to be activation or release of a latent degradative system which is specific for developing structures. However, the presence of such a system may not be necessary, for mature oral organelles constantly undergo protein turnover (Williams *et al.*, 1969). Recovery from puromycin, similar to that for cycloheximide, is not achieved (Frankel, 1970a).

The amino acid analog, *p*-fluorophenylalanine, has been studied for its effect on protein synthesis in *T. pyriformis*. This agent will block division of synchronized cells and cause resorption of developing oral primordia if it is added prior to 2 hr before division (Frankel, 1961, 1962; Rasmussen and Zeuthen, 1962, 1966a,b; Holz *et al.*, 1963). Its potency is increased with high levels of NaCl in the medium (Hoffman and Kramhøft, 1969); but its inhibition can be reversed by phenylalanine or methionine (Rasmussen and Zeuthen, 1966b), which may compete with the agent at the site of entry into the cell. *p*-Fluorophenylalanine can be incorporated into material which is precipitable with trichloroacetic acid (Rasmussen, 1968). Specifically, it is incorporated into ribosomal proteins; but ribosomes containing the analog have normal physico-chemical properties (Leick, 1969a). Some effects of the analog, perhaps secondary to its incorporation, are that it depresses uridine and thymidine uptake into nucleic acids (Rasmussen, 1968) and that it causes selective inhibition of the incorporation of inorganic phosphate into ribosomal ribonucleic acid, but not into transfer ribonucleic acid (Leick, 1968). To some degree, cells can become adapted to the presence of *p*-fluorophenylalanine in the growth medium (Rasmussen and Zeuthen, 1966b).

Four other amino acid analogs, ethionine, β-thienylalanine, azaserine, and canavanine, also show delay effects on synchronized division of *T. pyriformis* (Lee *et al.*, 1959; Holz *et al.*, 1963; Rasmussen and Zeuthen, 1966a). Their effects are not clearly understood, but it is known that ethionine is incorporated into the proteins of this organism (Gross and Tarver, 1955). A biochemical effect of canavanine is inhibition of incorporation of inorganic phosphate into ribosomal ribonucleic acid but not into transfer ribonucleic acid (Leick, 1968). The inhibition of cell division by azaserine is reversed by phenylalanine but not by glutamine (Lee and Yuzurihi, 1964). Azaserine has little effect on individual cell growth (Lee *et al.*, 1959).

When cells of *T. pyriformis* are transferred from a complete medium to one lacking amino acids, they cease dividing after several hours and begin to undergo oral replacement, a process that can be synchronized by a single, long exposure to heat (Frankel, 1969b, 1970b). The process can be blocked completely by actinomycin D, cycloheximide, or puromycin.

A toxic, undefined substance, fusarenon, isolated from rice grains infected with *Fusarium nivale* may have a specific effect on the protein synthesis of *T. pyriformis* (Nakano, 1968).

Division Protein

It is interesting that a number of the inhibitors of protein synthesis just mentioned set back the time of division if they are applied before a critical point in the interfission period. Applications after that point have no delaying effect. This transition point is found also with heat, 2,4-dinitrophenol, and anaerobiosis (Thormar, 1959; Hamburger, 1962; Rasmussen, 1963), colchicine (Wunderlich and Peyk, 1969; Nelsen, 1970), and high hydrostatic pressure (Simpson and Williams, 1970; Lowe-Jinde and Zimmerman, 1969); but protein synthesis of some kind appears to be a controlling feature in cell division. The data are consistent with the view that a certain level of a critical "division protein" (Rasmussen and Zeuthen, 1962) is necessary.

Attempts to isolate the "division protein" have been made (Watanabe and Ikeda, 1965a,b,c; Ikeda and Watanabe, 1965). The incorporation of ^{3}H-phenylalanine decreases during the heat treatment used to produce synchronous division, but rises again at the end of heat treatment to a rate that is one-tenth that of cells in the logarithmic phase of growth. This increase is also recognizable under starvation conditions, which would decrease the formation of proteins not involved in the forthcoming division. Thus, there is a strong possibility that this small protein fraction contains the division protein. By the use of pulse labeling and diethylaminoethyl (DEAE)-cellulose column chromatography, the division protein has been tentatively located in a particular fraction from the column. It is known that there are changes in the sulfhydryl content of the soluble portion of the cell during division (Scherbaum *et al.*, 1960; Katoh, 1960); and changes in incorporation of ^{3}H-methionine are noted mostly in the DEAE-cellulose fraction suspected of containing the division protein (Ikeda and Watanabe, 1965). Further characterization of this protein is difficult, and other workers have not found evidence for it (Lowe-Jinde and Zimmermann, 1971).

Other significant observations on division-related protein synthesis are those by Byfield and Scherbaum (1967c) and Byfield and Lee (1970a), who found that ribonucleic acid synthesis is approximately linear with time in growth-supporting media. Although others found some fluctuation in this rate during temperature-induced synchrony, the fluctuation is probably due to inadequate assays (Byfield and Lee, 1970a). If ribonucleic acid production is relatively constant and the synthesis of division-related proteins is drastically reduced by heat-shock treatment, a basis for the reduction could be the hydrolysis of template ribonucleic acid without concurrent translation for formation of division-related proteins. A temperature-induced denaturation of ribosomes has been noted. Ribosomes from two strains of *Tetrahymena pyriformis* are denatured at temperatures close to each strain's maximum growth temperature; both are considerably less stable than those of *Escherichia coli*. However, the same temperatures that initiate melting of isolated ribosomes also reduce the efficiency of messenger translation in intact cells (Byfield and Lee, 1969; Byfield *et al.*, 1969). Since ribosomes are not denatured by cold shocks which induce synchrony but the functioning of messenger ribonucleic acid is reduced, the defect introduced by temperature shifts is probably one involving translational efficiency (Byfield and Lee, 1970b,c). An earlier hypothesis that division-related proteins are denatured by heat treatment (Rasmussen and Zeuthen, 1962) is now less appealing.

Although it is certain that protein synthesis is inhibited at elevated temperatures (Byfield and Scherbaum, 1966, 1967a; Levy *et al.*, 1969), whether protein synthesis varies during the cycle of synchronous division following heat treatment is the subject of a dispute. Watanabe and Ikeda (1965a,b) made a strong point of this in their attempts to characterize the division protein. Plesner (1964) states that protein synthesis falls off just before division, and Hamburger and Zeuthen (1960) found a minimum rate of synthesis during fission. But Prescott (1960) and Crockett *et al.* (1965) found that the rate does not vary during synchronous division. The rate of protein synthesis is constant in the cell cycle of normal cells (Prescott, 1960) but drops in stationary-phase cells, which contain little or no rough endoplasmic reticulum (Elliott *et al.*, 1966).

Proteins of Cilia

The cross-sectional structure of a cilium from *Tetrahymena* is shown in Figs. 5.2 and 5.3 (Allen, 1967b). The two central fibers are surrounded by a central sheath, and spokes from this sheath connect with each of nine pairs of subfibers. One of each pair of subfibers has two arms protruding. The ciliary membrane surrounds the entire structure (Gibbons, 1963; Allen, 1968b). A recent investigation shows two previously undescribed rows of short projections on one of the central microtubules (Chasey, 1969). High-resolution electron micrographs reveal that the fibers of cilia of *Tetrahymena pyriformis* are composed of spherical subunits about 4 to 6 nm in diameter and arranged in straight longitudinal arrays to compose the walls of the fibers. When viewed in cross sections, each fiber appears to have thirteen subunits (Burnasheva *et al.*, 1968, 1969a). The basal bodies present in the oral apparatus are, like cilia, cylindrical with a ninefold symmetry (Wolfe, 1970). However, in basal bodies, central microtubules are absent; and the outer nine fibers are triplets (Figs. 5.4–5.7). The subfibers are composed of filamentous rows of globules with a spacing of 4.5 nm. One subfiber is only one-third the length of the basal body.

Chemical analyses of the cilia show that about 80% of the dry weight of these structures is protein (Watson *et al.*, 1961; Watson and Hopkins, 1962). Most of the remaining material is phospholipid, but there is 1–5% carbohydrate present (Watson and Hopkins, 1962; Culbertson, 1964, 1966a). The amino acid composition is not greatly different from that generally found in proteins (Watson and Hopkins, 1962; Watson *et al.*, 1964; Renaud *et al.*, 1968), although one group of investigators insists that hydroxyproline is present (Culbertson and Hull, 1963; Culbertson, 1966a). Wells (1960) also found hydroxyproline in a crude protein fraction of *T. pyriformis*, but no collagen or collagen-like protein can be demonstrated (Nordwig *et al.*, 1969). Guanine nucleotides are associated with the protein of the

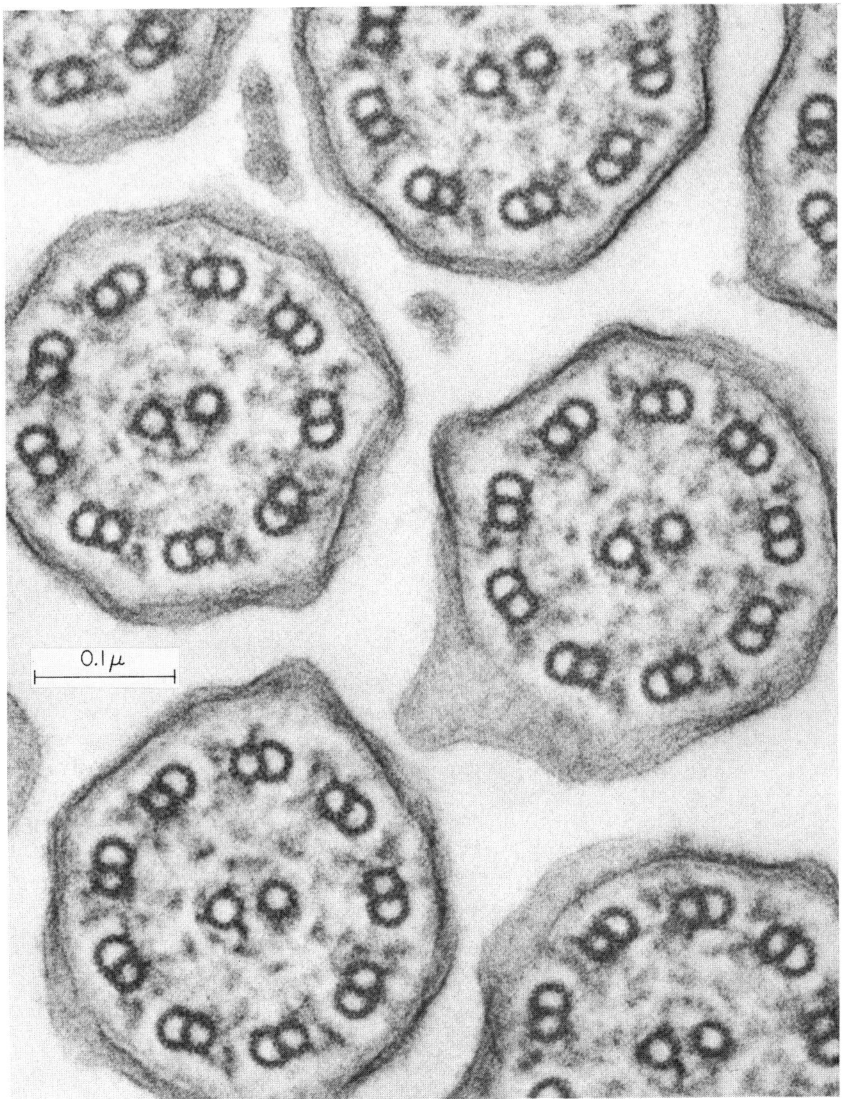

FIG. 5.2. Cross sections of cilia. Magnification: ×320,000. (Courtesy of Dr. I. L. Cameron and Mr. Glenn Williams.)

outer fibers of cilia in a ratio of 1 mole of nucleotide per protein subunit (Culbertson, 1966a; Stevens *et al.*, 1967).

The insolubility of cilia has hindered the study of the protein components of these structures. After dissolving cilia in dilute NaOH, Child (1959)

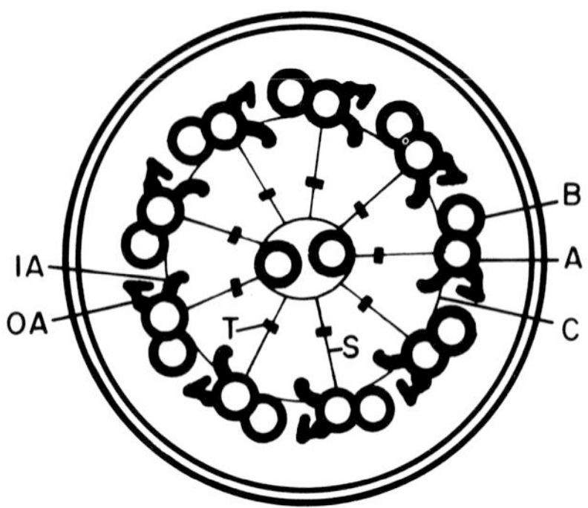

FIG. 5.3. Diagram of a cross section of a *Tetrahymena pyriformis* cilium. Abbreviations: A. subfiber A; B, subfiber B; C, connection between outer double fibers; OA, outer arm; IA, inner arm; S, spoke; T, thickened region along spike. (From Allen, 1968b.)

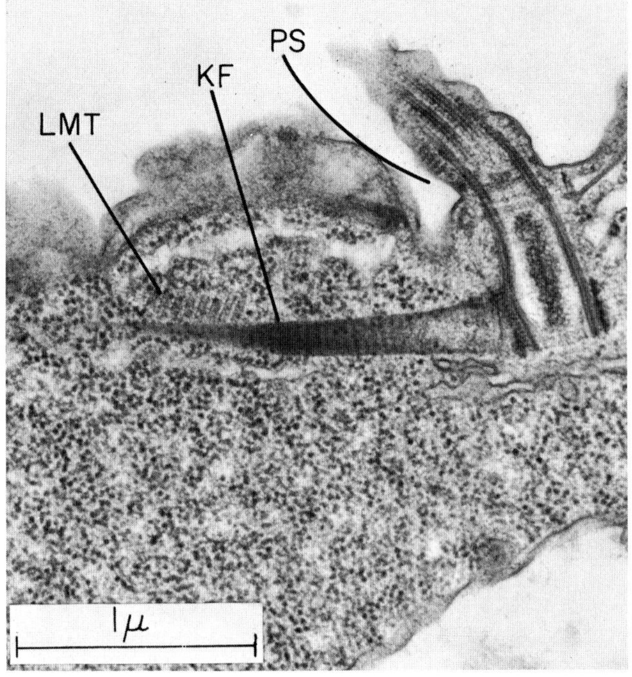

FIG. 5.4. Basal body of *Tetrahymena pyriformis* cilium. Kinetodesmal fiber (KF) shows periodicity. Longitudinal microtubules (LMT) and the perisomal sac (PS) are shown. Magnification: ×32,000. (Courtesy of Dr. I. L. Cameron and Mr. Glenn Williams.)

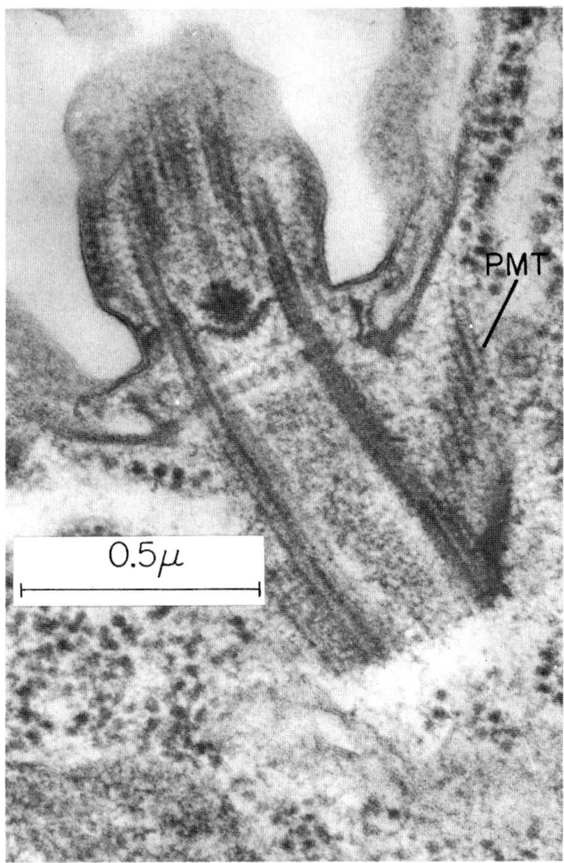

FIG. 5.5. Basal body showing posterior microtubules (PMT). Magnification: ×64,000. (Courtesy of Dr. I. L. Cameron and Mr. Glenn Williams.)

found only one boundary in the analytical ultracentrifuge. By extracting with salt solutions, other investigators (Burnasheva *et al.*, 1965) obtained three fractions. In an acid extract, Alexander *et al.* (1962) also found three components, one of which has an amino acid composition different from the total composition. Alexander (1965), using immunological techniques can identify eight to ten individual proteins.

An outstanding contribution was made with the discovery that the apparent insolubility of ciliary proteins is due to the presence of the membrane surrounding the cilia (Gibbons, 1963, 1965a). The membrane can be removed by digitonin; and the remaining material can be fractionated further

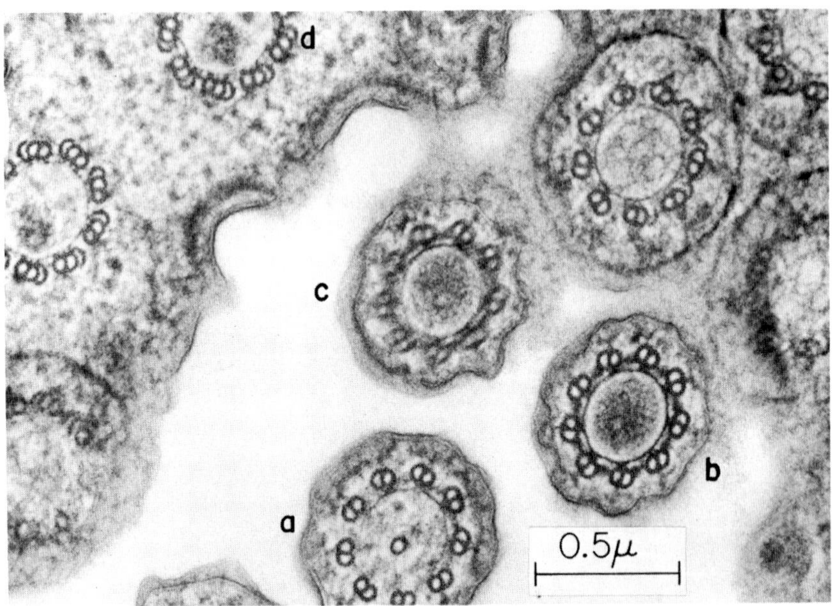

FIG. 5.6. Tangential section through basal bodies of cilia. Structures a, b, c, and d are progressively lower sections. Magnification: ×40,000. (Courtesy of Dr. I. L. Cameron and Mr. Glenn Williams.)

by dialysis against a solution of a chelator at low concentration. The soluble fraction (15% of the total ciliary protein) contains almost all of the ATPase activity. The insoluble fraction consists entirely of the outer fibers of the cilia; the central pair, the inner matrix, and the arms are removed. The structure can be partially restored by adding magnesium ions; and when this is done, ATPase activity returns to the fibers of the insoluble fraction. As seen in the electron microscope, the arms are present on the outer fibers after restoration. The conclusion is that ATPase resides in these arms. The soluble fraction contains protein components with sedimentation coefficients of 14 and 30 S (Gibbons and Rowe, 1965). The 30 S component appears to be a linear polymer of the globular 14 S subunit, which has a molecular weight of about 600,000 daltons. The 14 S material may represent the spherical subunits seen with the electron microscope (Burnasheva *et al.*, 1968, 1969a). The molecular weight of the 30 S structure is about 5.4×10^6 daltons. The protein present in this structure has been named "dynein," and it is the material that forms the arms on the outer fibers of the cilium. It has been implicated in the mechanism of ciliary motility, since its presence is required for sensitivity of light-scattering properties of cilia to ATP (Gibbons, 1964, 1965c).

The insoluble fraction, composed of the outer fibers, can be brought into solution by 0.6 *M* NaCl (Renaud *et al.*, 1968). Only one protein component, with a sedimentation coefficient of 4 S, is found. It has a molecular weight of 104,000 daltons in the presence of 5 *M* guanidine. This structure may be split into subunits with molecular weights of 55,000 daltons, for which there are about 7.5 sulfhydryl groups per subunit. After reducing and alkylating the disulfhydryl bonds in 8 *M* urea, the protein migrates as a single band on electrophoresis.

A study of ATP on the hydrodynamic properties of cilia (Raff and Blum, 1966) led to an investigation (Raff and Blum, 1969) which apparently confirms much of the data of Gibbons and Rowe (1965). At low concentrations of ATP, glycerinated cilia swell; but at high concentrations, partial dissolution occurs (Raff and Blum, 1969). The soluble proteins thus obtained show three major peaks on sucrose-density gradient centrifugation: a 4 S peak with no ATPase activity; a 14 S peak with low ATPase; and a 30 S peak with high ATPase. The pellet remaining after extraction contains ciliary membranes and outer fiber pairs. The 30 S protein can recombine functionally with the pellet and can be removed again by ATP.

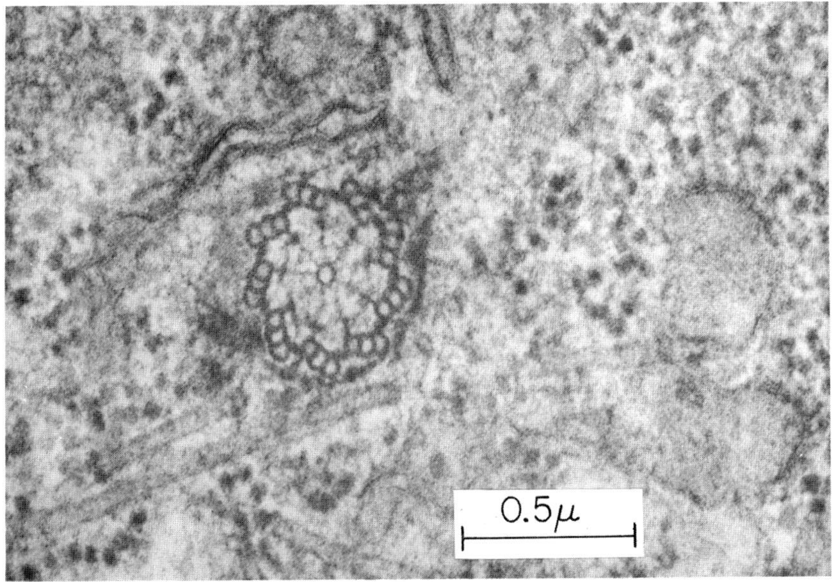

Fig. 5.7. Cartwheel base with adjacent basal tubules. Magnification: ×48,000. (Courtesy of Dr. I. L. Cameron and Mr. Glenn Williams.)

Watson *et al.* (1964) used 3.1 *N* acetic acid to dissolve the ciliary proteins. In the ultracentrifuge, the extract showed three components with sedimentation coefficients of 1.7, 4, and 9 S. Following precipitation with trichloroacetic acid, the 4 S component can be extracted with ethanol. This group has been unable to find subunits attributable to dynein, but they too recommend that disulfide bonds be reduced and alkylated prior to separation of proteins (Watson and Hynes, 1966). The material insoluble in 3.1 *N* acetic acid has a high proportion of phosphorus, probably representing the phospholipids of the ciliary membrane (Silvester, 1964).

The outstanding enzymatic activity of cilia is ATPase (Child, 1959; Gibbons, 1963, 1966; Warnock and van Eys, 1963; Burnasheva *et al.*, 1963); adenosine diphosphate (ADP) and adenosine monophosphate (AMP) are also hydrolyzed (Child, 1959; Warnock and van Eys, 1963; Burnasheva *et al.*, 1969b). Adenylate kinase (Culbertson, 1966b), arginine kinase (Watts and Bannister, 1970), a nonspecific cholinesterase (Schuster and Hershenov, 1969), and a nonspecific phosphatase cleaving α-glycerophosphate and inorganic pyrophosphate (Burnasheva *et al.*, 1969b) are other enzymes demonstrated in cilia.

The ciliary ATPase requires a divalent cation for activity (Burnasheva *et al.*, 1963; Burnasheva and Raskidnaya, 1968; Gibbons, 1966). Several ions may be utilized; but magnesium may be physiologically active, since it is required for the motility of glycerinated cilia (Gibbons, 1965b; Winicur, 1967). Both ATP and 2′-deoxyATP are substrates, and the pH optimum is between 8.5 and 9.0 for both. The Michaelis constant for ATP cleavage by 14 S dynein is 3.5×10^{-5} *M*; that for 30 S dynein is 1.1×10^{-5} *M*.

The ciliary ATPase of *T. pyriformis* can be used to study the effects of tobacco smoke constituents, some of which are known to be ciliastatic in mammals (Culbertson, 1968). Their toxicity may be due to destruction of mitochondria, which supply energy to the cilia (Kennedy and Elliott, 1970).

Although inhibitors of protein synthesis block cilia regeneration, a more selective inhibitor is colchicine (Rosenbaum and Carlson, 1969), which exerts its effect without inhibiting either ribonucleic acid or protein biosynthesis. A suggested action of colchicine is that it interferes with the assembly of ciliary subunit proteins.

Characterization of Nonprotein Nitrogen

Single cells of *Tetrahymena pyriformis* contain about 2 ng of protein nitrogen (3% of the dry weight) (Singer and Eiler, 1960; Wu and Hogg, 1952) and about 2.5 ng of nonprotein nitrogen (4% of the dry weight) (Wu and Hogg, 1952), of which free amino acids are outstanding. These constitute 12.6–23.2% of the total amino acid content (Wu and Hogg, 1952; Scherbaum *et al.*, 1959; Christensson, 1959). The free amino acids and related compounds have been characterized by use of microbiological assays (Wu and Hogg, 1952), by use of paper chromatography (Wu and Hogg, 1956; Christensson, 1959; Schleicher, 1959; Scherbaum *et al.*, 1959; Wells, 1960; Loefer and Scherbaum, 1958, 1961, 1963), and by use of automatic amino acid analyzers (Hill and van Eys, 1964; Wragg *et al.*, 1965). Although many different strains of *T. pyriformis* and a number of other species have been examined, the amino acids reported have not varied greatly. Alanine, arginine, aspartic acid, asparagine, cystine, glutamic acid, glutamine, glycine, histidine, leucine, isoleucine, methionine, phenylalanine, lysine, proline, serine, threonine, tyrosine, and valine are consistently found; on occasion, glutathione (Scherbaum *et al.*, 1959), ornithine (Wu and Hogg, 1956; Loefer and Scherbaum, 1961), and tryptophan (Wu and Hogg, 1956) have been reported. Although the qualitative results do not differ greatly, quantitatively great variations are seen. However, the variability of the amino acid pool can be accounted for. The concentration of amino acids in the medium can strongly influence the intracellular concentration, and the size of the pool varies with the stage of growth (Hill and van Eys, 1964). Quantitative variation in extractable amino acid pools also accompany differences in the kind and amount of carbohydrate supplied in the growth medium (Reynolds and Wragg, 1962; Reynolds, 1970).

There is evidence for a considerable pool of peptides in *T. pyriformis.* The amino acid nitrogen which is soluble in trichloroacetic acid accounts for one-third of the total in this organism (Wu and Hogg, 1952); and the free and total nonprotein α-amino nitrogen values are 18.3 and 26.7%, respectively, of the total nitrogen of the cells (Wu and Hogg, 1956). Conjugated amino acids, or peptides, account for the difference between these values; but an extensive search revealed that no discrete peptides could be found (Hill and van Eys, 1964). Apparently those present represent intermediates in the formation of protein, and there is no peptide present in isolable quantity.

Arginine, Proline, and the Absence of a Urea Cycle

Arginine is required by *Tetrahymena* (Hogg and Elliott, 1951). In the work to establish the arginine requirement, it was reported that the organism could be adapted to synthesize small amounts of arginine by culturing it in a medium containing the other essential amino acids and that citrulline and ornithine plus asparagine could replace the arginine requirement (Elliott and Hogg, 1952). Later evidence (Hill and van Eys, 1965; Hill and Chambers, 1967) makes a pathway leading from ornithine to arginine seem unlikely. It is conceivable that there is a novel pathway in adapted cells leading to the formation of arginine, but it is much more likely that the amino acids replacing arginine contained small amounts of arginine as a contaminant. The D-isomer of arginine is equally as effective as the L-isomer in promoting growth (Hogg and Elliott, 1951).

Early reports regarding the presence of a urea cycle in *T. pyriformis* were conflicting. Nardone and Wilbur (1950) and Seaman (1954) reported that urea could be found in cultures of this organism. Seaman (1954) further reported that all of the enzymes associated with the urea cycle were present and later produced cytochemical evidence for the presence of urease (Seaman, 1959). However, there is now overwhelming evidence to show that there is no active urea cycle and that *Tetrahymena* is ammonotelic (Lwoff and Roukelmann, 1926; Doyle and Harding, 1937; Wu and Hogg, 1952; Dewey *et al.*, 1957; Hill and van Eys, 1965; Hill and Chambers, 1967). Dewey *et al.* (1957), using both enzyme assays and growth studies, extensively examined several strains for their ability to accomplish the reactions of the urea cycle. Although there was a breakdown of arginine, no urea cycle enzymes could be detected. They did find that neither citrulline, ornithine, nor proline could replace arginine in the medium but that these compounds did have pronounced sparing effects on the requirement for arginine. This indicated the existence of a pathway whereby arginine is converted to these compounds. It is now known that the pathway from arginine to proline proceeds as follows (Hill and van Eys, 1965; Hill and Chambers, 1967):

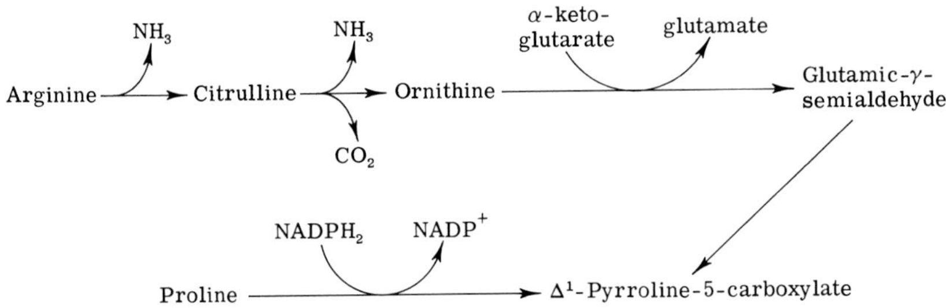

The first step is catalyzed by arginine deiminase, and this enzyme has been purified seventy-five-fold. The enzyme has a pH optimum of 9.3 and is located in the soluble portion of the cell. L-Ornithine is a competitive inhibitor with a K_i of 1.08×10^{-5} M, compared to a K_m for L-arginine of 2.84×10^{-3} M. Thus, the first portion of the pathway is controlled by feedback inhibition of the deiminase by ornithine, which explains the sparing effects of citrulline and ornithine on the arginine requirement.

The second enzyme, citrulline hydrolase, is a novel one, being reported only for this organism. It has a pH optimum of 6.8, is soluble, and has a substrate concentration at half-maximum velocity of 1.06×10^{-3} M. It was purified seventy-nine-fold. The kinetics are unusual in that a reciprocal plot of velocity and substrate gives a curved line. The apparent order of the reaction can be determined by use of the equation proposed by Hill (1910) to describe the kinetics of hemoglobin oxygenation. This equation as applied to citrulline hydrolase is $\log v/(V - v) = \log K + n \log[\text{citrulline}]$, where v = velocity, V = maximum velocity, n = order of reaction, and K = apparent association constant. The value for n is 1.41. Similar data for other enzymes have been interpreted to mean that there is more than one binding site for substrate on the enzyme and that there is "cooperativity" between the sites, the value for n being a measure of the interaction (Monod *et al.*, 1963). Even though this enzyme has unusual kinetics, it serves no known function in the regulation of the pathway.

Ornithine δ-transaminase accomplishes the next step of the series. The enzyme has been purified 105-fold and the pH optimum is 9.0. When cells are broken by filtration, the activity is associated with particles from the cells. The Michaelis constants for α-ketoglutarate and L-ornithine are 9.1×10^{-3} and 1.9×10^{-3} M, respectively. L-Valine is a weak competitive inhibitor of ornithine with a K_i of 4.2×10^{-3} M. Pyridoxal phosphate is necessary for the reaction.

Following spontaneous cyclization of glutamate γ-semialdehyde to Δ^1-pyrroline-5-carboxylate, the formation of proline proceeds with a reductive step involving this compound, reduced nicotinamide adenine dinucleotide phosphate ($NADPH_2$), and Δ^1-pyrroline-5-carboxylate reductase. The pH optimum is 9.1; and the Michaelis constants are, for L-Δ^1-pyrroline-5-carboxylate, 1.8×10^{-3} M and, for $NADPH_2$, 3.2×10^{-5} M. This enzyme has been purified thirty-eight-fold.

Tracer studies confirm that these enzymes are active in the intact cell (Hill and Chambers, 1967).

The conversion of ornithine to proline by *T. pyriformis* proceeds by a pathway identical to that in *Neurospora crassa, Escherichia coli*, and mammals (Yura and Vogel, 1959; Smith and Greenberg, 1957; Peraino and Pitot, 1963; Vogel and Kopac, 1959; Baich and Pierson, 1965; Reed and Lukens, 1966); but the production of ornithine is unique in that arginine deiminase and citrulline hydrolase are involved. Deiminase is known to be present in *Streptococcus* (Slade, 1953), *Mycoplasma* (Barile *et al.*, 1966), and *Crithidia* (Kidder *et al.*, 1966).

An interesting report regarding arginine metabolism in serine-antagonized cells has appeared (Wragg *et al.*, 1965). When excess serine is added to the medium, the conversion of arginine to ornithine is increased. The authors postulate that the excess serine inhibits some energy-yielding reactions in the cell and that the cell resorts to the breakdown of arginine to provide energy. However, it does not seem possible that the purely hydrolytic reactions catalyzed by arginine deiminase and citrulline hydrolase could yield any usable energy (Hill and van Eys, 1965). It is possible that serine stimulates the deiminase directly; this point has not been investigated.

The Question of a Phosphagen

The term "phosphagen" refers to a compound that serves as a reservoir of labile phosphate and participates in phosphoryl group exchange with adenine nucleotides. *N*-Phosphorylcreatine (Eggleton and Eggleton, 1927) is present in all types of vertebrate muscle; and *N*-phosphorylarginine, isolated by Meyerhof and Lohman (1928), is present in the tissues of many invertebrates. A number of other phosphagens have been isolated from various sources (Thoai *et al.*, 1953a, b, 1963; Robin and Thoai, 1962; Roche *et al.*, 1956).

The inorganic polyphosphates present in yeast (Hoffmann-Ostenhof *et al.*, 1954; Yoshida and Yamataka, 1953), certain bacteria (Kornberg *et al.*, 1956), algae (Schmidt and King, 1961), and phytoflagellates (Reichenow, 1928) appear to function as reservoirs of labile phosphate for the synthesis of ATP; but whether these materials should be considered phosphagens is in dispute (Hoffmann-Ostenhof and Weigert, 1952; Ennor and Morrison, 1958). No inorganic polyphosphate is present in *Tetrahymena pyriformis* (Hill *et al.*, 1965).

"Volutin" granules which contain salts of inorganic pyrophosphate are found in *T. pyriformis*; the pyrophosphate might serve as a reservoir of energy (Rosenberg, 1966; Munk and Rosenberg, 1969; Rosenberg and Munk, 1969). The granules are deposited when normal growth is retarded through some nutritional deficiency or when the stationary phase of growth is reached. The presence of two divalent cations is required. Calcium and magnesium are most effective, but strontium can substitute for either. During phosphate deprivation, the granules disappear rapidly by the action of an inorganic pyrophosphatase associated with the granules. Since no phosphotransferase activity can be demonstrated, the authors conclude that the deposited pyrophosphate functions in phosphate storage rather than in phosphoryl group transfer.

In one report regarding the presence of a phosphagen in *T. pyriformis*, data are presented to show that a phosphagen is present and that it differs from *N*-phosphorylcreatine and *N*-phosphorylarginine (Seaman, 1952). However, the evidence is meager. Grossly impure preparations were used and conclusions were based on the rate of liberation of inorganic phosphate by mild acid hydrolysis and on the so-called molybdate retardation factor. The liberation of inorganic phosphate under conditions of low pH is not necessarily indicative of the presence of a phosphagen, as other compounds liberate inorganic phosphate under similar conditions. The effect of molybdate is to lower the rate of hydrolysis of some *N*-phosphorylated compounds under acid conditions. This effect has been found to be dependent on acid strength and on the molybdate-phosphagen ratio (Griffiths, 1958). Variation in these factors can account for the large variations in the values reported for the retardation factor for *N*-phosphorylarginine. The values range from 1.5 to 30 (Ennor and Morrison, 1958). Production of inorganic phosphate by extracts of *T. pyriformis* under mild acid conditions can be attributed to hydrolysis of nucleoside triphosphates (Hill *et al.*, 1965).

Other studies present evidence for the presence of *N*-phosphorylarginine in *T. pyriformis* (Dorner, 1963; Kidder, 1967). After hydrolysis of the suspected phosphagen, arginine can be detected chromatographically; and the eluted guanidine supports the growth of *T. pyriformis* in an arginine-free medium. The most convincing evidence concerning a phosphagen from *T. pyriformis* is the report that the enzyme ATP–arginine phosphotransferase is active (Robin and Viala, 1966). The investigators who made this discovery agree with Hill *et al.* (1965) that arginine is the principal guanidine in the cells, but neither group was able to isolate or characterize a phosphagen. It is possible that *N*-phosphorylarginine is only an intermediate in the production of a compound which serves as a phosphate reservoir. Another possibility is that the *N*-phosphorylarginine is compartmentized in the cell, perhaps in the kinetosomes, so that its overall concentration in the cell is very low. In this regard, a recent report shows that arginine kinase is located predominantly in the cilia (Watts and Bannister, 1970). A necessary conclusion, from all the evidence, is that *N*-phosphorylarginine and arginine kinase play some role in the storage of labile phosphate.

Serine and Threonine

L-Serine was at first thought to be an essential amino acid for growth of *Tetrahymena* (Kidder and Dewey, 1947), but subsequent studies (Dewey and Kidder, 1960a) revealed that this is true for only one of the sixteen strains tested. The requirement for serine observed earlier was due to to deficiencies of folic acid and glycine in the medium. The unnatural isomer, D-serine, is inhibitory to growth (Kidder and Dewey, 1951).

The growth response of this organism to varying amounts of glycine provides a method for studying the biosynthesis of serine (Dewey and Kidder, 1960a). When the folic acid content of the medium is low, thus preventing appreciable synthesis of serine from glycine by means of the enzyme serine hydroxymethyltransferase, growth is severely retarded. The addition of larger amounts of folic acid makes better growth possible. When glucose is added to the medium without increasing the folic acid content, growth is almost entirely suppressed. On the other hand, if both glucose and folic acid are added without increasing the amount of glycine in the medium, there is a very large stimulatory effect which is much greater than that obtained upon addition of glucose to a medium containing serine. Therefore, the addition of glucose alone produces only a small stimulation by its serving as an energy source. Dewey and Kidder (1960a) view these results as indicating that glucose serves as a precursor of serine and, subsequently, of glycine. Fructose, maltose, mannose, pyruvate, lactate, fatty acids containing an odd number of carbon atoms, and low levels of ethanol and acetaldehyde show similar stimulatory effects and are also considered as precursors of serine and glycine (Dewey and Kidder, 1960b). These workers postulated that all of these compounds are converted to 3-phosphoglycerate and, subsequently, to serine through the intermediates phosphohydroxypyruvate and phosphoserine (Pizer, 1964).

Glycine, cysteine, ethanolamine, and thymine show sparing effects for the requirement for serine (Dewey and Kidder, 1960b), an indication that serine contributes to their formation. Since the synthesis of glycine is accomplished by the enzyme serine hydroxymethyltransferase, this amino acid would be expected to spare the requirement for serine. Cysteine presumably exerts a sparing effect because its postulated precursor, cystathionine, is formed from serine. Ethanolamine is the decarboxylation product of serine and could eliminate the serine requirement for its synthesis. Thymine synthesis is known to require a 1-carbon fragment. Such a fragment is produced in the glycine conversion; and the presence of thymine would eliminate this requirement.

Excess serine in the medium is inhibitory to growth (Singer, 1961; Wragg *et al.*, 1965), but the inhibition may be reversed by the addition of more arginine (Wragg *et al.*, 1965), more methionine (Singer, 1961) or, in the presence of dextran, by glucose (Reynolds, 1969).

L-Threonine is required by *Tetrahymena.* It cannot be replaced by the D-isomer, by DL-allothreonine, by α-amino-*n*-butyrate or by homoserine

(Kidder and Dewey, 1951; Kidder, 1967). Growth studies with sixteen strains of *T. pyriformis* show that fifteen of these are able to use threonine for synthesis of serine if sufficient folic acid is present (Dewey and Kidder, 1960a). The intermediate in this case is thought to be glycine, which is formed from threonine by the enzyme threonine aldolase (Dewey and Kidder, 1960b). The authors think that one strain lacks threonine aldolase, since it can synthesize serine from added glycine but not from added threonine; but this notion is now open to question, for serine hydroxymethyltransferase and threonine aldolase were reported to be the same enzyme in rabbit liver (Schirch and Gross, 1968).

Glycine and Porphyrins

Tetrahymena pyriformis does not require exogenous glycine for growth. Instead, glycine is an excretory product (Reynolds, 1970). The derivatives *N*-acetyl-, *N*-propionyl-, and *N*-butyrylglycine have no effect; but *N*-laurylglycine is inhibitory (Kidder and Dewey, 1951).

Tetrahymena vorax and *T. pyriformis* are known to accomplish the biosynthesis of porphyrins, utilizing glycine and δ-aminolevulinic acid as precursors (Rudzinska and Granick, 1953; Lascelles, 1957). δ-Aminolevulinate dehydrase, which converts δ-aminolevulinate to porphobilinogen, is present (Granick, 1954). The porphyrins formed are protoporphyrin, coproporphyrin III, and uroporphyrin III. A requirement for pyridoxal and pantothenate for the synthesis of these compounds is noted (Lascelles, 1957).

Methionine and Cysteine

The requirement by *Tetrahymena pyriformis* for L-methionine can be spared by cystine, homocystine, or D-methionine (Kidder and Dewey, 1951; Elliott and Hogg, 1952). Choline and vitamin B_{12}, which are involved in 1-carbon metabolism, are inactive in sparing the requirement. On the basis of the sparing effects, it seems probable that methionine is converted to cysteine via the following pathway:

$$\text{Methionine} \rightarrow \text{homocysteine} \xrightarrow{\text{serine}} \text{cystathionine} \rightarrow \text{cysteine} + \text{homoserine}$$

Enzymes accomplishing these reactions are present in a number of other organisms (Meister, 1957). *N*-Acetyl-, *N*-propionyl-, and *N*-butyryl-DL-methionine are weakly active in replacing methionine, but *N*-lauryl-DL-methionine is inhibitory, as is the analog DL-ethionine (Kidder and Dewey, 1951). The inhibition by ethionine can be reversed by methionine.

The uptake of L-methionine by *T. pyriformis* follows Michaelis–Menten kinetics, with a K_m of 4.1 μm (Hoffman and Rasmussen, 1969). L-Phenylalanine inhibits the influx of L-methionine, which suggests that a common carrier operates for the two amino acids.

A complex relationship between methionine and serine is observed when excesses of these compounds are used to inhibit growth (Singer, 1961). Inhibition due to excess serine is relieved by raising the level of methionine or by adding pyrimethamine, an agent which lowers the effective folic acid concentration. Inhibition due to excess methionine is relieved by supplying more serine or adding ethionine, and the inhibition is intensified by adding pyrimethamine. Inhibition by excess methionine plus serine is reversed by increasing the threonine concentration. The biochemical basis for these relationships remains unknown.

Methionine and cysteine protect *T. pyriformis* from the effects of ultraviolet irradiation of the medium (Sullivan, 1959; Sullivan *et al.*, 1962).

Cysteine is not an essential amino acid; and other than its sparing effect on the requirement for methionine (Kidder and Dewey, 1951), nothing is known about its metabolism in *Tetrahymena*.

Valine, Leucine, and Isoleucine

The valine requirement by *Tetrahymena pyriformis* is absolute (Kidder, 1967). In an early study valine appeared to be synthesized slowly (Kidder and Dewey, 1947), but the amino acids used in the medium undoubtedly contained a small amount of valine. α-Ketoisovalerate can be substituted for valine (Kidder, 1967), although much higher amounts are required. The D-isomer of valine is inactive, as is the hydroxy analog.

Two enzymes involved in valine catabolism, β-hydroxyisobutyryl CoA deacylase and β-hydroxyisobutyrate dehydrogenase, are present in *T. pyriformis* (Redinda and Coon, 1957; Robinson and Coon, 1957). Apparently the pathway for valine breakdown,

$$\text{Valine} \rightarrow \rightarrow \text{isobutyryl CoA} \rightarrow \rightarrow \rightarrow \beta\text{-hydroxyisobutryate} \rightarrow \rightarrow \rightarrow \text{succinyl CoA}$$

is the same as for other organisms (Meister, 1957).

Another absolute requirement for *Tetrahymena* is L-leucine (Kidder and Dewey, 1947). The D-isomer in this case does not stimulate growth and is actually inhibitory (Kidder and Dewey, 1951). Kidder (1967) reported that the hydroxy analog of isoleucine, α-hydroxyisocaproate, cannot replace leucine nor spare the requirement for it; but Hogg and Elliott (1951) and Kidder (1967) state that the keto analog, α-ketoisocaproate, can replace leucine. *N*-Acetyl-L-leucine and *N*-propionyl-L-leucine can also, to a small extent, replace the leucine requirement (Kidder and Dewey, 1947). *N*-Butyryl-L-leucine spares the requirement, but *N*-lauryl- and *N*-salicyl-L-leucine are inhibitory. There is evidence for a mechanism for induction of leucine uptake by *T. pyriformis* (Schaeffer and Dunham, 1970). The system is induced only by strongly hydrophobic, neutral amino acids. The Michaelis constant for uptake is not changed in the induced system; but the maximum velocity increases greatly, suggesting that the number of leucine transport sites is increased. Cycloheximide, chloramphenicol, and actinomycin D inhibit the induction.

L-Isoleucine is also an absolute requirement for growth (Kidder and Dewey, 1947). Neither the D-isomer nor the hydroxy analog, β-methyl-α-hydroxyvaleric acid, is active. However, Hogg and Elliott (1951) state that the keto analog, β-methyl-α-ketovaleric acid, can substitute for leucine.

Lysine, Alanine, Glutamic Acid, and Histidine

L-Lysine is a required amino acid for *Tetrahymena pyriformis*; only the L-isomer is active (Kidder and Dewey, 1951). This amino acid is the only one of the essential group that is not inhibitory when the concentration in the medium is increased severalfold (Dewey and Kidder, 1958). However, lysine at high concentration can delay the onset of cell division of synchronized cells (Rasmussen and Zeuthen, 1966a). The effect is stronger at high levels of NaCl (Hoffman and Kramhøft, 1969); but several amino acids, when added with lysine, can release the inhibition.

Alanine is not required for growth by *T. pyriformis*; and it may not be extensively degraded, for it is an excretory product (Reynolds, 1970). Up to 90% of the ninhydrin-positive material excreted by synchronized cells in an inorganic medium is alanine (Cann, 1968).

Glutamic acid is also not required, but it is both excreted (Reynolds, 1970) and actively metabolized. Glutamate dehydrogenase, a ubiquitous enzyme, produces α-ketoglutarate, an intermediate in the Krebs cycle (Roth and Eichel, 1955; Thompson and Wingo, 1959) and a substrate for transamination reactions with alanine and aspartic acid (Müller, 1969). On gel electrophoresis of cell homogenates, multiple molecular forms of this enzyme are seen (Allen, 1968). A study of glutamate dehydrogenase activity at different stages of growth shows that it is the same when expressed on the basis of nitrogen content, although mature cultures contain more activity per cell (Thompson and Wingo, 1959). α-Ketoglutarate may also be produced by glutamate–oxalacetate transaminase, which is present in the organism (Wingo and Thompson, 1960).

L-Histidine is an absolute requirement for growth of *Tetrahymena*

(Kidder and Dewey, 1947). The amount required for optimal growth is small in dilute media but is increased in more concentrated media. The D-isomer of histidine is inactive, and imidazole does not spare the amino acid. Hogg and Elliott (1951) reported that carnosine (β-alanyl-L-histidine) can replace histidine as a growth requirement and that imidazololactate spares the histidine requirement.

Tryptophan and Serotonin

L-Tryptophan is also required by this organism (Kidder and Dewey, 1947). The D-isomer is inactive (Kidder and Dewey, 1951), as are the keto analog of tryptophan, indole-3-pyruvic acid, and *N*-acetyltryptophan. However, Seaman (1955) reported that tryptophan can be transaminated by extracts of *Tetrahymena pyriformis* in the presence of α-ketoglutarate. Indole, a product of tryptophan degradation, is toxic to the organism (Hogg and Elliott, 1951); it induces lysis of a number of ciliated Protozoa (Bailey and Russell, 1965). Growth inhibition is also achieved by use of the tryptophan analog 7-azatryptophan (Kidder and Dewey, 1955); but L-tryptophan reverses the inhibition. A surprising report has appeared regarding the presence of serotonin (5-hydroxytryptamine) in this organism and in the protozoan *Crithidia fasiculata* (Janakadevi *et al.*, 1966b). This compound, in mammals, is a potent vasoconstrictor found in several tissues, but its function in *Tetrahymena* is open to speculation. The serotonin content of cells is maximal during stationary phase (Brizzi and Blum, 1970). It is increased by addition of 5-hydroxytryptophan to the medium, but L-tryptophan, reserpine, *p*-chlorophenylalanine, and demethylimipramine do not have this effect.

Phenylalanine and Tyrosine

L-Phenylalanine is another requirement for *Tetrahymena pyriformis*. Only the L-isomer is active, but the D-isomer is not inhibitory (Kidder and Dewey, 1951). Tyrosine is effective in sparing phenylalanine, indicating that phenylalanine is the precursor of tyrosine, as it is in other organisms. Kidder (1967) and Hogg and Elliott (1951) found that the keto analog can replace the requirement for phenylalanine, although higher amounts are required. Tyrosine ethyl ester is readily utilized (Kidder, 1967).

The kinetics of the uptake of phenylalanine by *T. pyriformis* has been studied (Stephens and Kerr, 1962a, b; Hoffman and Rasmussen, 1969). At low concentrations the uptake of phenylalanine follows Lineweaver–

Burk kinetics with a Michaelis constant of 0.12 mM, but the maximum velocity for this process is exceeded at higher levels. Entrance of phenylalanine into the cell is specific for the L-isomer, is inhibited by 2,4-dinitrophenol, and occurs against a concentration gradient, indicating that a membrane transport system involving a mobile carrier is operative. The same carrier may operate for L-methionine, since this amino acid inhibits phenylalanine uptake. The addition of L-phenylalanine to a culture decreases the intracellular sodium content, suggesting that this ion has a role in the uptake of the amino acid (Hoffman and Kramhøft, 1969).

When *T. pyriformis* is exposed to peptides containing phenylalanine or tyrosine, oxygen consumption is increased more than for the free amino acid (Roth and Eichel, 1961). However, reaction products of phenylalanine metabolism are not oxidized, nor are derivatives of phenylalanine or tyrosine in which the amino group is blocked. Such data are not easily explained, but perhaps transport across the cell membrane is involved.

Tyrosine aminotransferase is present in *T. pyriformis* (Maurides and D'Iorio, 1969; Porter and Blum, 1970). During all growth phases, the activity is repressed by glucose; but acetate represses the enzyme only in the logarithmic phase (Maurides and D'Iorio, 1969).

The phenylalanine analog, *p*-fluorophenylalanine, has an effect on morphogenesis (Frankel, 1961), on protein synthesis, and on cell division (Cerroni and Zeuthen, 1962), as described in the section on protein synthesis. The *ortho* and *meta* isomers are also inhibitory to growth (Kidder, 1967). The inhibition by the *ortho* isomer can be reversed by tyrosine, and that by β-thienylalanine, another phenylalanine analog, can be reversed by tyrosine or phenylalanine (Kidder and Dewey, 1951). *p*-Chlorophenylalanine is not toxic (Brizzi and Blum, 1970).

Ubiquinones

There has been some confusion regarding the pathway for biosynthesis of the ubiquinones which are present in *Tetrahymena* (Taketomi, 1961; Crane, 1962; Vakirtzi-Lemonias *et al.*, 1963). Rats use the preformed aromatic rings of phenylalanine for the manufacture of ubiquinones, but two separate investigations agree that this is not the case for *T. pyriformis* (Braun *et al.*, 1963; Miller, 1965). Another possible mechanism for ring formation, the condensation of acetate units, has been ruled out (Braun *et al.*, 1963). Braun *et al.* (1963) also rule out the shikimic acid pathway, but Miller (1965) presents decisive evidence to show that the pathway is indeed operative. The conclusions of Braun *et al.* (1963) were derived from experiments showing that unlabeled shikimate did not dilute the labeling of ubiquinones by radioactive glucose. Miller (1965) used a lower concentration of glucose and increased the amount of shikimate. He also degraded the ubiquinones to show that the ring of ubiquinones was labeled by glucose, shikimate, and *p*-hydroxybenzoate.

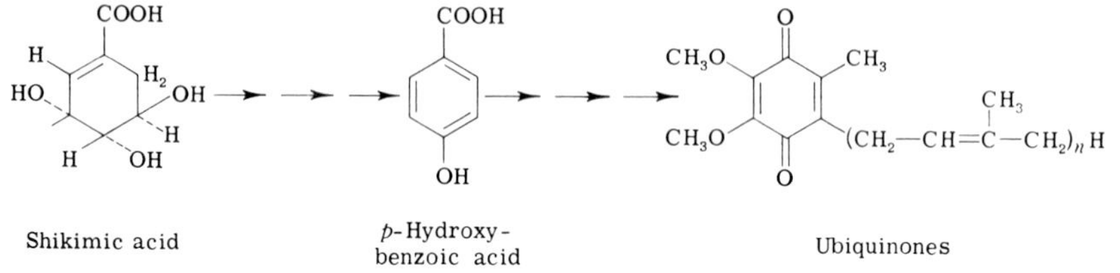

Shikimic acid → → → *p*-Hydroxybenzoic acid → → → Ubiquinones

Although this organism requires phenylalanine for growth, it retains the ability to synthesize aromatic rings as do bacteria. The shikimate pathway may have arisen from evolutionary changes involving deletion of enzymes in the terminal portion of the pathway of phenylalanine and tryptophan biosynthesis.

Catecholamines

A surprising development in a closely related field is the discovery that *Tetrahymena pyriformis* can make the catecholamines, epinephrine and norepinephrine (Janakidevi *et al.*, 1966a). These compounds, which are familiar as hormones mainly of the adrenal medulla of mammals, also occur in invertebrates of the phyla Aschelminthes, Annelida, Arthropoda, Mollusca, and Echinodermata (Von Euler, 1961). In addition to their property of raising the blood pressure of mammals, epinephrine also increases blood glucose by promoting glycogenolysis in muscle. It acts directly on the enzyme adenyl cyclase, which forms 3′,5′-cyclic AMP (Sutherland and Rall, 1960). Cyclic AMP activates phosphorylase kinase, which, in turn, promotes conversion of inactive glycogen phosphorylase b to active phosphorylase a. Although enzymatic data are lacking, it is possible that epinephrine has a similar effect in *T. pyriformis*, for methylxanthines inhibiting the phosphodiesterase which converts 3′,5′-cyclic AMP to AMP also inhibit the growth of this organism (Blum, 1967). In mammals the catecholamines also increase the blood level of nonesterified fatty acids by stimulating their release from adipose tissue (Butcher *et al.*, 1965), an action also mediated by 3′,5′-cyclic AMP. Whether these substances produce an elevation of free fatty acids in the ciliates remains to be determined, as well as the question whether there are present prostaglandins to counteract such an effect (Butcher and Baird, 1968).

The biosynthesis of the catecholamines by *T. pyriformis* presumably proceeds as it does in mammals (Scheme 5.1) (Axelrod *et al.*, 1958), since radioactivity of phenylalanine, tyrosine, and dihydroxyphenylalanine is incorporated readily into the amines (Janakidevi *et al.*, 1966a).

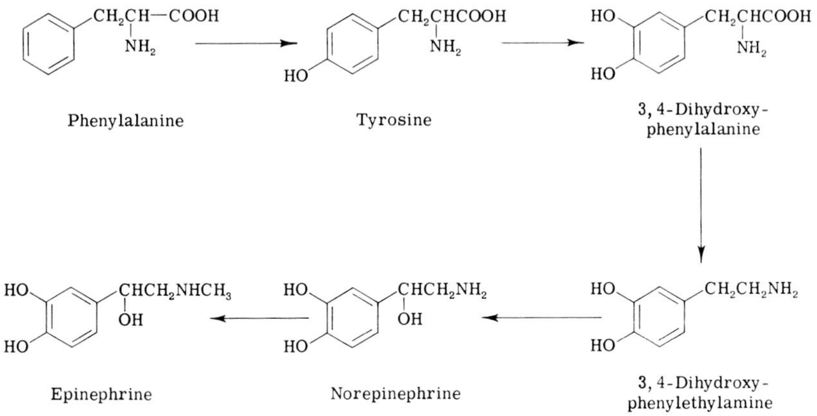

$CH_2CH(NH_2)-COOH$ → $CH_2CH(NH_2)COOH$ (HO) → HO, HO $CH_2CH(NH_2)COOH$

Phenylalanine → Tyrosine → 3, 4-Dihydroxyphenylalanine

↓

HO, HO $CH_2CH_2NH_2$ → HO, HO $CH(OH)CH_2NH_2$ → HO, HO $CH(OH)CH_2NHCH_3$

3, 4-Dihydroxyphenylethylamine → Norepinephrine → Epinephrine

Scheme 5.1

Reserpine, a drug which diminishes the activity of the central nervous system in man, inhibits the growth of *T. pyriformis* and reduces the cellular content of the catecholamines (Blum *et al.*, 1966). The effect of reserpine is partially reversed by glucose and by low concentrations of tranylcypromine, a monoamine oxidase inhibitor (Blum, 1967). Growth of the organism is also sensitive to certain classic agents that block α- or β-adrenergic sites in mammals (Blum, 1967; Iwata *et al.*, 1967, 1969). Those that are effective are dibenzyline, propanolol, and dichloroisoproterenol. These agents have an effect on the glycogen content consistent with the view that an adrenergic metabolic control system is operative in *Tetrahymena*. Further evidence for such a system comes from the fact that a known inhibitor of monoaminase oxidase, α-methylphenethylhydrazine, also inhibits the growth of *T. pyriformis* (Blum, 1967).

REFERENCES

Alexander, J. B. (1965). *In* "Progress in Protozoology," *Proc. 2nd Int. Conf. Protozool., London,* 1965.

Alexander, J. B., Silvester, N. R., and Watson, M. R. (1962). *Biochem. J.* **85,** 27P.

Allen, R. D. (1968a). *J. Protozool.* **15,** Suppl. 7.

Allen, R. D. (1968b). *J. Cell Biol.* **37,** 825.

Allen, S. L. (1968). *Ann. N.Y. Acad. Sci.* **151,** 190.

Anderson, M. E., and Williams, H. H. (1951). *J. Nutr.* **44,** 335.

Argetsinger, J. (1965). *J. Cell Biol.* **24,** 154.

Arlock, P., Heby, O., and Holm, B. (1969). *J. Protozool.* **16,** Suppl. 33.

Axelrod, J., Senoh, S., and Witkop, B. (1958). *J. Biol. Chem.* **233,** 697.

Baich, A., and Pierson, D. J. (1965). *Biochim. Biophys. Acta* **104,** 397.

Bailey, R. W., and Russell, G. B. (1965). *Nature (London)* **208,** 1001.

Barile, M. F., Schimke, R. T., and Riggs, D. B. (1966). *J. Bacteriol.* **91,** 189.

Baudhuin, P., Müller, M., Poole, B., and de Duve, C. (1965). *Biochem. Biophys. Res. Commun.* **20,** 53.

Bergner, H., München, H., and Koch, R. (1968). *Arch. Tierernaehr.* **18,** 212.

Blum, J. J. (1967). *Proc. Nat. Acad. Sci. U. S.* **58,** 81.

Blum, J. J., Kirschner, N., and Utley, J. (1966). *Mol. Pharmacol.* **2,** 606.

Bortle, L., and Oleson, J. J. (1954–55). *Antibiot. Annu.* 770.

Boyne, A. W., Price, S. A., Rosen, G. D., and Stott, J. A. (1967). *Brit. J. Nutr.* **21,** 181.

Braun, R., Dewey, V. C., and Kidder, G. W. (1963). *Biochemistry* **2,** 1070.

Brizzi, G., and Blum, J. J. (1970). *J. Protozool.* **17,** 553.

Burnasheva, S. A., and Raskidnaya, N. V. (1968). *Dokl. Akad. Nauk. SSSR* **179,** 719.

Burnasheva, S. A., Efremenko, M. V., and Lyubimova, M. N. (1963). *Biokhimiya* **28,** 547.

Burnasheva, S. A., Efremenko, M. V., Chumakova, L. P., and Zeuva, L. V. (1965). *Biokhimiya* **30,** 765.

Burnasheva, S. A., Yurzina, G. A., and Beskrovnova, M. V. (1968). *Tsitologiya* **10,** 249.

Burnasheva, S. A., Yurzina, G. A., and Lyubimova, M. N. (1969a). *Tsitologiya* **11,** 695.

Burnasheva, S. A., Daiya, D. Y., and Yurzina, G. A. (1969b). *Biokhimiya* **34,** 443.

Butcher, R. W., and Baird, C. E. (1968). *J. Biol. Chem.* **243,** 1731.

Butcher, R. W., Ho, R. J., Meng, H. C., and Sutherland, E. W. (1965). *J. Biol. Chem.* **240,** 4515.

Byfield, J. E., and Lee, Y. C. (1969). *J. Protozool.* **16,** Suppl. 14.

Byfield, J. E., and Lee, Y. C. (1970a). *J. Protozool.* **17,** 445.

Byfield, J. E., and Lee, Y. C. (1970b). *J. Protozool.* **17,** Suppl. 19.

Byfield, J. E., and Lee, Y. C. (1970c). *Exp. Cell Res.* **61,** 42.

Byfield, J. E., and Scherbaum, O. H. (1966). *Life Sci.* **5,** 2263.

Byfield, J. E., and Scherbaum, O. H. (1967a). *Proc. Nat. Acad. Sci. U.S.* **57,** 602.

Byfield, J. E., and Scherbaum, O. H. (1967b). *Science* **156,** 1504.

Byfield, J. E., and Scherbaum, O. H. (1967c). *J. Cell. Physiol.* **70,** 265.

Byfield, J. E., Lee, Y. C., and Bennett, L. R. (1969). *Biochem. Biophys. Res. Commun.* **37,** 806.

Cameron, I. L., Cline, G. B., Padilla, G. M., Miller, O. L., Jr., and van Dreal, P. A. (1966). *Nat. Cancer Inst. Monogr.* **21,** 361.

Cann, J. R. (1968). *C. R. Trav. Lab. Carlsburg* **36,** 319.

Celliers, P. G. (1961). *S. Afr. J. Agr. Sci.* **4,** 191.

Cerroni, R. E., and Zeuthen, E. (1962). *Exp. Cell Res.* **26,** 604.

Chasey, D. (1969). *J. Cell Sci.* **5,** 453.

Chatton, E., and Lwoff, A. (1935). *C. R. Acad. Sci. Ser.* **D118,** 1068.

Chi, J. C. H., and Suyama, Y. (1970). *Fed. Proc. Fed. Amer. Soc. Exp. Biol.* **29,** 866.

Child, F. M. (1959). *Exp. Cell Res.* **18,** 258.

Child, F. M. (1965). *J. Cell Biol.* **27,** 18A.

Christensson, E. (1959). *Acta Physiol. Scand.* **45,** 339.

Crane, F. L. (1962). *Biochemistry* **1,** 510.

Crockett, R. L., Dunham, P. B., and Rasmussen, L. (1965). *C.R. Trav. Lab. Carlsberg* **34,** 451.

Culbertson, J. R. (1964). *J. Protozool.* **11,** Suppl. 22.
Culbertson, J. R. (1966a). *J. Protozool.* **13,** 397.
Culbertson, J. R. (1966b). *Science* **153,** 1390.
Culbertson, J. R. (1968). *J. Protozool.* **15,** Suppl. 15.
Culbertson, J. R., and Hull, R. W. (1963). *J. Protozool.* **10,** Suppl. 8.
Dewey, V. C. (1967). In "Chemical Zoology" (M. Florkin and B. T. Scheer, eds.), Vol. 1, p. 161. Academic Press, New York.
Dewey, V. C., and Kidder, G. W. (1958). *Arch. Biochem. Biophys.* **73,** 29.
Dewey, V. C., and Kidder, G. W. (1960a). *J. Gen. Microbiol.* **22,** 72.
Dewey, V. C., and Kidder, G. W. (1960b). *J. Gen. Microbiol.* **22,** 79.
Dewey, V. C., and Kidder, G. W. (1964). *Biochem. Pharmacol.* **13,** 353.
Dewey, V. C., Heinrich, M. R., and Kidder, G. W. (1957). *J. Protozool.* **4,** 211.
Dickie, N., and Liener, I. E. (1962a). *Biochim. Biophys. Acta* **64,** 41.
Dickie, N., and Liener, I. E. (1962b). *Biochim. Biophys. Acta* **64,** 52.
Dorner, D. B. (1963). M. S. thesis. Amherst College, Amherst, Mass.
Doyle, W. L., and Harding, J. R. (1937). *J. Exp. Biol.* **14,** 462.
Dunn, M. S., and Rockland, L. B. (1947). *Proc. Soc. Exp. Biol. Med.* **64,** 377.
Eggleton, P., and Eggleton, G. P. (1927). *Biochem. J.* **21,** 190.
Elliott, A. M. (1959). *Annu. Rev. Microbiol.* **13,** 79.
Elliott, A. M., and Clark, G. M. (1958a). *J. Protozool.* **5,** 235.
Elliott, A. M., and Clark, G. M. (1958b). *J. Protozool.* **5,** 240.
Elliott, A. M., and Hayes, R. E. (1955). *J. Protozool.* **2,** Suppl. 8.
Elliott, A. M., and Hogg, J. F. (1952). *Physiol. Zool.* **25,** 318.
Elliott, A. M., Addison, M. A., and Carey, S. E. (1962). *J. Protozool.* **9,** 135.
Elliott, A. M., Studier, M. A., and Work, J. A. (1964). *J. Protozool.* **11,** 370.
Elliott, A. M., Travis, D. M., and Work, J. A. (1966). *J. Exp. Zool.* **161,** 177.
Ennor, A. H., and Morrison, J. F. (1958). *Physiol. Rev.* **38,** 631.
Fernell, W. R., and Rosen, G. D. (1956). *Brit. J. Nutr.* **10,** 143.
Frankel, J. (1961). *J. Protozool.* **8,** Suppl. 12.
Frankel, J. (1962). *C.R. Trav. Lab. Carlsberg* **33,** 1.
Frankel, J. (1967). *J. Cell Biol.* **34,** 841.
Frankel, J. (1969a). *J. Cell. Physiol.* **74,** 135.
Frankel, J. (1969b). *J. Protozool.* **16,** 26.
Frankel, J. (1970a). *J. Cell. Physiol.* **76,** 55.
Frankel, J. (1970b). *J. Exp. Zool.* **173,** 79.
Gavin, R. H., and Frankel, J. (1969). *J. Cell. Physiol.* **74,** 123.
Gibbons, I. R. (1963). *Proc. Nat. Acad. Sci. U.S.* **50,** 1002.
Gibbons, I. R. (1964). *J. Cell. Biol.* **23,** 35A.
Gibbons, I. R. (1965a). *Arch. Biol.* **76,** 317.
Gibbons, I. R. (1965b). *J. Cell Biol.* **25,** 400.
Gibbons, I. R. (1965c). *J. Cell Biol.* **26,** 707.
Gibbons, I. R. (1966). *J. Biol. Chem.* **241,** 5590.
Gibbons, I. R., and Rowe, A. J. (1965). *Science* **149,** 424.
Gorovsky, M. A. (1969). *J. Cell Biol.* **43,** 46A.
Granick, S. (1954). *Science* **120,** 1105.
Griffiths, D. E. (1958). Ph.D. thesis, Australian National University.
Gross, D., and Tarver, H. (1955). *J. Biol. Chem.* **217,** 169.
Hamburger, R. (1962). *C. R. Trav. Lab. Carlsberg* **32,** 359.
Hamburger, R., and Zeuthen, E. (1960). *C. R. Trav. Lab. Carlsberg* **32,** 1.

Hartman, H., and Dowben, R. M. (1970a). *J. Cell Biol.* **45,** 676.
Hartman, H., and Dowben, R. M. (1970b). *Biochem. Biophys. Res. Commun.* **40,** 964.
Hermolin, J., and Zimmerman, A. M. (1969). *Cytobios* **1,** 247.
Hill, A. V. (1910). *J. Physiol.* (*London*) **40,** iv.
Hill, D. L., and Chambers, P. (1967). *Biochim. Biophys. Acta* **148,** 435.
Hill, D. L., and van Eys, J. (1964). *J. Tenn. Acad. Sci.* **39,** 117.
Hill, D. L., and van Eys, J. (1965). *J. Protozool.* **12,** 259.
Hill, D. L., Judd, J., and van Eys, J. (1965). *Comp. Biochem. Physiol.* **14,** 1.
Hoffman, E. J. (1965). *J. Cell Biol.* **25,** 217.
Hoffmann, E. K., and Kramhøft, B. (1969). *Exp. Cell Res.* **56,** 265.
Hoffmann, E., and Rasmussen, L. (1969). *J. Protozool.* **16,** Suppl. 33.
Hoffmann-Ostenhof, O., and Weigert, W. (1952). *Naturwissenschaften* **39,** 303.
Hoffmann-Ostenhof, O., Kenedy, J., Keck, K., Gabriel, O., and Schonfellinger, H. W. (1954). *Biochim. Biophys. Acta* **14,** 285.
Hogg, J. F., and Elliott, A. M. (1951). *J. Biol. Chem.* **192,** 131.
Holz, G. G., Jr. (1964). *In* "Biochemistry and Physiology of Protozoa" (S. H. Hutner, ed.), Vol. 3, p. 199. Academic Press, New York.
Holz, G. G., Jr., Erwin, J. A., and Wagner, B. (1961). *J. Protozool.* **8,** 297.
Holz, G. G., Jr., Erwin, J., Wagner, B., and Rosenbaum, N. (1962). *J. Protozool.* **9,** 359.
Holz, G. G., Jr., Rasmussen, L., and Zeuthen, E. (1963). *C. R. Trav. Lab. Carlsberg* **33,** 289.
Hurwitz, J., Furth, J. J., Maloney, M., and Alexander, M. (1962). *Proc. Nat. Acad. Sci. U.S.* **48,** 1222.
Ikeda, M., and Watanabe, Y. (1965). *Exp. Cell Res.* **39,** 584.
Iwata, H., Kariya, K., and Fugimoto, S. (1967). *Jap. J. Pharmacol.* **17,** 328.
Iwata, H., Kariya, K., and Fugimoto, S. (1969). *Jap. J. Pharmacol.* **19,** 275.
Janakidevi, K., Dewey, V. C., and Kidder, G. W. (1966a). *J. Biol. Chem.* **241,** 2576.
Janakidevi, K., Dewey, V. C., and Kidder, G. W. (1966b). *Arch. Biochem. Biophys.* **113,** 758.
Kamath, J. K., and Ambegakar, S. D. (1968). *J. Nutr. Diet.* **5,** 197.
Katoh, A. K. (1960). *J. Protozool.* **7,** Suppl. 21.
Kennedy, J. R., and Elliott, A. M. (1970). *Science* **168,** 1096.
Kessler, D. (1961). M.S. thesis, Syracuse University, Syracuse, New York.
Kidder, G. W. (1967). *In* "Chemical Zoology" (M. Florkin and B. T. Scheer, eds.), Vol. 1, p. 93. Academic Press, New York.
Kidder, G. W., and Dewey, V. C. (1947). *Proc. Nat. Acad. Sci. U.S.* **33,** 347.
Kidder, G. W., and Dewey, V. C. (1951). *In* "Biochemistry and Physiology of Protozoa" (A. Lwoff, ed.), Vol. 1, p. 323. Academic Press, New York.
Kidder, G. W., and Dewey, V. C. (1955). *Biochim. Biophys. Acta* **17,** 288.
Kidder, G. W., Davis, J. S., and Cousens, K. (1966). *Biochem. Biophys. Res. Commun.* **24,** 365.
Kimball, R. F. (1964). *In* "Biochemistry and Physiology of Protozoa" (S. H. Hutner, ed.), Vol. 3, p. 244. Academic Press, New York.
Kornberg, A., Kornberg, S. R., and Sims, E. S. (1956). *Biochim. Biophys. Acta* **20,** 215.
Kumar, A. (1968). *J. Cell Biol.* **39,** 76A.
Kumar, A. (1970). *J. Cell Biol.* **45,** 623.
Lantos, T., Müller, M., Törö, I., Druga, A., and Vargha, I. (1964). *Acta Biol.* (*Budapest*) **6,** Suppl. 22.
Lascelles, J. (1957). *Biochem. J.* **66,** 65.

Lawrie, N. R. (1937). *Biochem. J.* **31,** 789.
Lee, K-H., and Yuzuriha, Y. (1964). *J. Pharm. Sci.* **53,** 290.
Lee, K-H., Yuzuriha, Y., and Eiler, J. J. (1959). *J. Amer. Pharm. Ass. Pract. Pharm. Ed.* **48,** 470.
Leick, V. (1968). *J. Protozool.* **15,** Suppl. 37.
Leick, V. (1969a). *Eur. J. Biochem.* **8,** 215.
Leick, V. (1969b). *Eur. J. Biochem.* **8,** 221.
Leick, V., and Plesner, P. (1968a). *Biochim. Biophys. Acta* **169,** 398.
Leick, V., and Plesner, P. (1968b). *Biochim. Biophys. Acta* **169,** 409.
Leick, V., Engberg, J., and Emmersen, J. (1970). *Eur. J. Biochem.* **13,** 238.
Leon, S. A., and Mahler, H. R. (1968). *Arch. Biochem. Biophys.* **126,** 305.
Letts, P. J., and Zimmerman, A. M. (1970). *J. Protozool.* **17,** 593.
Levy, M. R., and Elliott, A. M. (1968). *J. Protozool.* **15,** 208.
Levy, M. R., Gollon, C. E., and Elliott, A. M. (1969). *Exp. Cell Res.* **55,** 295.
Loefer, J. B., and Scherbaum, O. H. (1958). *Anat. Record* **131,** 576.
Loefer, J. B., and Scherbaum, O. H. (1961). *J. Protozool.* **8,** 184.
Loefer, J. B., and Scherbaum, O. H. (1963). *J. Protozool.* **10,** 275.
Lowe-Jinde, L., and Zimmerman, A. M. (1969). *J. Protozool.* **16,** 226.
Lowe-Jinde, L., and Zimmerman, A. M. (1971). *J. Protozool.* **18,** 20.
Lwoff, A. (1932). "Recherches biochimiques sur la nutrition des Protozoaires, le pouvoir de synthèse." Masson, Paris.
Lwoff, A., and Roukhelman, N. (1926). *C. R. Acad. Sci. Ser.* **D183,** 156.
Lyttleton, J. W. (1963). *Exp. Cell Res.* **31,** 385.
Mager, J. (1960). *Biochim. Biophys. Acta* **38,** 150.
Mager, J., and Lipmann, F. (1958). *Proc. Nat. Acad. Sci. U. S.* **44,** 305.
Martin, T. E. (1968). *J. Cell Biol.* **39,** 85A.
Martin, T. E. (1969). *J. Cell Biol.* **43,** 85A.
Martin, T. E., and Wool, I. G. (1969). *J. Mol. Biol.* **43,** 151.
Maurides, C., and D'Iorio, A. (1969). *Biochem. Biophys. Res. Commun.* **35,** 467.
Mefferd, R. B., Jr., Matney, T. S., and Loefer, J. B. (1952). *Proc. Soc. Protozool.* **3, 17.**
Meister, A. (1957). "Biochemistry of the Amino Acids." Academic Press, New York.
Meyerhoff, O., and Lohmann, K. (1928). *Biochem. Z.* **196,** 22.
Miller, J. E. (1965). *Biochim. Biophys. Res. Commun.* **19,** 335.
Moner, J. G. (1964). *J. Protozool.* **11,** Suppl. 34.
Monod, J., Changeux, J-P., and Jacob, F. (1963). *J. Mol. Biol.* **6,** 306.
Müller, M. (1969). *Ann. N.Y. Acad. Sci.* **168,** 292.
Müller, M., Hogg, J. F., and de Duve, C. (1968). *J. Biol. Chem.* **243,** 5385.
Müller, W., and Crothers, D. M. (1968). *J. Mol. Biol.* **35,** 251.
Munk, N., and Rosenberg, H. (1969). *Biochim. Biophys. Acta* **177,** 629.
Nachtwey, D. S., and Dickinson, W. J. (1967). *Exp. Cell Res.* **47,** 581.
Nakano, N. (1968). *Jap. J. Med. Sci. Biol.* **21,** 351.
Nardone, R. M., and Wilbur, C. G. (1950). *Proc. Soc. Exp. Biol. Med.* **75,** 559.
Nathans, D. (1964). *Fed. Proc. Fed. Amer. Sci. Exp. Biol.* **23,** 984.
Nelsen, E. M. (1970). *J. Exp. Zool.* **175,** 69.
Nordwig, A., Hayduk, U., and Gerish, G. (1969). *Hoppe-Seyler's Z. Physiol. Chem.* **350,** 245.
Peraino, C., and Pitot, H. C. (1963). *Biochim. Biophys. Acta* **73,** 222.
Perry, R. P. (1967). *Progr. Nucl. Acid Res. Mol. Biol.* **6,** 219.
Pilcher, H. L., and Williams, H. H. (1954). *J. Nutr.* **53,** 589.

Pitelka, D. R., and Child, F. M. (1964). *In* "Biochemistry and Physiology of Protozoa" (S. H. Hutner, ed.), Vol. 3, p. 131. Academic Press, New York.
Pizer, L. I. (1964). *J. Biol. Chem.* **239,** 4219.
Plesner, P. (1961). *Cold Spring Harbor Symp. Quant. Biol.* **26,** 159.
Plesner, P. (1964). *C. R. Trav. Lab. Carlsberg* **34,** 1.
Porter, P., and Blum, J. J. (1970). *J. Protozool.* **17,** Suppl. 12.
Prescott, D. M. (1960). *Annu. Rev. Physiol.* **22,** 17.
Raff, E. C., and Blum, J. J. (1966). *J. Cell Biol.* **31,** 445.
Raff, E. C., and Blum, J. J. (1969). *J. Biol. Chem.* **244,** 366.
Rampton, V. W. (1962). *Nature (London)* **195,** 195.
Rasmussen, L. (1963). *C. R. Trav. Lab. Carlsberg* **33,** 53.
Rasmussen, L. (1968). *J. Protozool.* **15,** Suppl. 37.
Rasmussen, L., and Zeuthen, E. (1962). *C. R. Trav. Lab. Carlsberg* **32,** 333.
Rasmussen, L., and Zeuthen, E. (1966a). *C. R. Trav. Lab. Carlsberg* **35,** 85.
Rasmussen, L., and Zeuthen, E. (1966b). *Exp. Cell Res.* **41,** 462.
Redinda, G., and Coon, M. J. (1957). *J. Biol. Chem.* **225,** 523.
Reed, D. E., and Lukens, L. N. (1966). *J. Biol. Chem.* **241,** 264.
Reichenow, E. (1928). *Arch. Protistenk.* **61,** 144.
Renaud, F. L., Rowe, A. J., and Gibbons, I. R. (1968). *J. Cell Biol.* **36,** 79.
Reynolds, H. (1969). *J. Protozool.* **16,** 204.
Reynolds, H. (1970). *J. Bacteriol.* **104,** 719.
Reynolds, H., and Wragg, J. B. (1962). *J. Protozool.* **9,** 214.
Robin, Y., and Thoai, N. V. (1962). *Biochim. Biophys. Acta* **63,** 481.
Robin, Y., and Viala, B. (1966). *Comp. Biochem. Physiol.* **18,** 405.
Robinson, W. G., and Coon, M. J. (1957). *J. Biol. Chem.* **225,** 511.
Roche, J., Thoai, N. V., Robin, Y., and Pradel, L. A. (1956). *C. R. M. Acad. Sci.* **150,** 1684.
Rockland, L. B., and Dunn, M. S. (1949). *Food Technol.* **3,** 289.
Rosen, G. D., and Fernell, W. R. (1956). *Brit. J. Nutr.* **10,** 156.
Rosenbaum, J. L., and Carlson, K. (1969). *J. Cell Biol.* **40,** 415.
Rosenbaum, J. L., and Holz, G. G., Jr. (1966). *J. Protozool.* **13,** 115.
Rosenberg, H. (1966). *Exp. Cell Res.* **41,** 397.
Rosenberg, H., and Munk, N. (1969). *Biochim. Biophys. Acta* **184,** 191.
Roth, J. S., and Eichel, H. J. (1955). *Biol. Bull.* **108,** 308.
Roth, J. S., and Eichel, H. J. (1961). *J. Protozool.* **8,** 69.
Roth, J. S., Eichel, H. J., and Ginter, E. (1954). *Arch. Biochem. Biophys.* **48,** 112.
Rudzinska, M. A., and Granick, S. (1953). *Proc. Soc. Exp. Biol. Med.* **83,** 525.
Satir, B., and Rosenbaum, J. L. (1965). *J. Protozool.* **12,** 397.
Schaeffer, J. F., and Dunham, P. B. (1970). *J. Protozool.* **17,** Suppl. 20.
Scherbaum, O. H. (1963). *In* "The Cell in Mitosis" (L. Levine, ed.), p. 125. Academic Press, New York.
Scherbaum, O. H., James, T. W., and Jahn, T. L. (1959). *J. Cell. Comp. Physiol.* **53,** 119.
Scherbaum, O. H., Louderback, A. L., and Brown, A. (1960). *J. Protozool.* **7,** Suppl. 25.
Schirch, L., and Gross, T. (1968). *J. Biol. Chem.* **243,** 5651.
Schleicher, J. D. (1959). *Cath. Univ. Amer. Biol. Stud.* **53,** 1.
Schmidt, R. R., and King, K. W. (1961). *Biochim. Biophys. Acta* **47,** 391.
Schuster, F. L., and Hershenov, B. (1969). *Exp. Cell Res.* **55,** 385.
Seaman, G. R. (1952). *Biochim. Biophys. Acta* **9,** 693.
Seaman, G. R. (1954). *J. Protozool.* **1,** 207.

Seaman, G. R. (1955). *In* "Biochemistry and Physiology of Protozoa" (S. H. Hutner and A. Lwoff, eds.), Vol. 2, p. 91. Academic Press, New York.
Seaman, G. R. (1959). *J. Protozool.* **6,** 331.
Seaman, G. R. (1960). *Exp. Cell Res.* **21,** 292.
Seaman, G. R. (1962). *Biochim. Biophys. Acta* **55,** 889.
Shaw, R. F., and Williams, N. E. (1963). *J. Protozool.* **10,** 486.
Silvester, N. R. (1964). *J. Mol. Biol.* **8,** 11.
Simpson, R. E., and Williams, N. E. (1970). *J. Exp. Zool.* **175,** 85.
Singer, S. (1961). *J. Protozool.* **8,** 265.
Singer, W., and Eiler, J. J. (1960). *J. Amer. Pharm. Ass. Sci. Ed.* **49,** 669.
Slade, H. D. (1953). *Arch. Biochem. Biophys.* **42,** 204.
Smith, M. E., and Greenberg, D. E. (1957). *J. Biol. Chem.* **226,** 317.
Stephens, G. C., and Kerr, N. S. (1962a). *Am. Zool.* **2,** 450.
Stephens, G. C., and Kerr, N. S. (1962b). *Nature (London)* **194,** 1094.
Stevens, R. E., Renaud, F. L., and Gibbons, I. R. (1967). *Science* **156,** 1606.
Stott, J. A., and Smith, H. (1966). *Brit. J. Nutr.* **20,** 663.
Stott, J. A., Smith, H., and Rosen, G. D. (1963). *Brit. J. Nutr.* **17,** 227.
Sullivan, W. D. (1959). *Trans. Amer. Microsc. Soc.* **78,** 181.
Sullivan, W. D., McCormick, S. J., and McCormick, E. A. (1962). *Trans. Amer. Microsc. Soc.* **81,** 80.
Sutherland, E. W., and Rall, T. W. (1960). *Pharmacol. Rev.* **12,** 265.
Suyama, Y., and Eyer, J. (1967). *Biochem. Biophys. Res. Commun.* **28,** 746.
Swift, H., Adams, B. J., and Larsen, K. (1964). *J. Roy. Microsc. Soc.* **83,** 161.
Taketomi, T. (1961). *Z. Allg. Mikrobiol.* **1,** 331.
Teunisson, D. J. (1961). *Anal. Biochem.* **2,** 405.
Thoai, N. V., Roche, J., Robin, Y., and Thiem, N. V. (1953a). *Biochim. Biophys. Acta* **11,** 593.
Thoai, N. V., Roche, J., Robin, Y., and Thiem, N. V. (1953b). *C. R. M. Acad. Sci.* **147,** 1670.
Thoai, N. V., di Jeso, F., and Robin, Y. (1963). *C. R. Acad. Sci. Ser.* **D256,** 4525.
Thompson, W. D., and Wingo, W. J. (1959). *J. Cell Comp. Physiol.* **53,** 413.
Thormar, H. (1959). *C. R. Trav. Lab. Carlsberg* **31,** 207.
Turner, G., and Lloyd, D. (1970). *Biochem. J.* **116,** 41P.
Vakirtzi-Lemonias, C., Kidder, G. W., and Dewey, V. C. (1963). *Comp. Biochem. Physiol.* **8,** 331.
Viswanatha, T., and Liener, I. E. (1955). *Arch. Biochem. Biophys.* **56,** 222.
Viswanatha, T., and Liener, I. E. (1956). *Arch. Biochem. Biophys.* **61,** 410.
Vogel, R. H., and Kopac, M. J. (1960). *Biochim. Biophys. Acta* **37,** 539.
von Euler, U. S. (1961). *Nature (London)* **190,** 170.
Warnock, L. G., and van Eys, J. (1963). *J. Cell. Comp. Physiol.* **61,** 309.
Watanabe, Y., and Ikeda, M. (1965a). *Exp. Cell Res.* **38,** 432.
Watanabe, Y., and Ikeda, M. (1965b). *Exp. Cell Res.* **39,** 443.
Watanabe, Y., and Ikeda, M. (1965c). *Exp. Cell Res.* **39,** 464.
Watson, M. R., and Hopkins, J. M. (1962). *Exp. Cell Res.* **28,** 280.
Watson, M. R., and Hynes, R. D. (1966). *Exp. Cell Res.* **42,** 348.
Watson, M. R., Hopkins, J. M., and Randall, J. T. (1961). *Exp. Cell Res.* **23,** 629.
Watson, M. R., Alexander, J. B., and Silvester, N. R. (1964). *Exp. Cell Res.* **33,** 112.
Watts, D. C., and Bannister, L. H. (1970). *Nature (London)* **226,** 450.
Weller, D. L., Raina, A., and Johnstone, B. (1968). *Biochim. Biophys. Acta* **157,** 558.

Wells, C. (1960). *J. Protozool.* **7,** 7.
Whitson, G. L., Padilla, G. M., and Fisher, W. D. (1966). *Exp. Cell Res.* **42,** 438.
Williams, N. E., Michelson, O., and Zeuthen, E. (1969). *J. Cell Sci.* **5,** 143.
Wingo, W. J., and Thompson, W. D. (1960). *Fed. Proc. Fed. Amer. Exp. Sci. Biol.* **19,** 243.
Winicur, S. (1967). *J. Cell Biol.* **35,** C7.
Wolfe, J. (1970). *J. Cell Sci.* **6,** 679.
Wragg, J. B., Reynolds, H., and Pelczar, M. J., Jr. (1965). *J. Bacteriol.* **90,** 748.
Wu, C., and Hogg, J. F. (1952). *J. Biol. Chem.* **198,** 753.
Wu, C., and Hogg, J. F. (1956). *Arch. Biochem. Biophys.* **62,** 70.
Wunderlich, F., and Peyk, D. (1969). *Experentia* **25,** 1278.
Yoshida, A., and Yamataka, A. (1953). *J. Biochem.* **40,** 85.
Yura, T., and Vogel, H. J. (1959). *J. Biol. Chem.* **234,** 335.

CHAPTER 6

Purine, Pyrimidine, and Nucleic Acid Metabolism

Purine and Pyrimidine Requirements

Tetrahymena pyriformis was the first organism shown to require an exogenous purine and pyrimidine for growth (Kidder and Dewey, 1945), but it is now apparent that a number of ciliated protozoans share such a requirement. It is true that mutagenic agents can produce purine and pyrimidine requirements in molds and bacteria, but these requirements are usually associated with the loss of a single enzyme. In *T. pyriformis*, evidently all of the enzymes of the *de novo* pathway leading to the production of inosine monophosphate (IMP) are missing, along with those converting IMP to guanosine monophosphate (GMP).

Guanine is an absolute requirement for the organism, although it can be supplied either as the free base, the nucleoside, or the nucleotide. Adenine and hypoxanthine spare the guanine requirement, but xanthine and uric acid are inert (Kidder and Dewey, 1948, 1951; Kidder *et al.*, 1950;

Holz, 1964). By use of guanine-8-^{14}C and adenine-8-^{14}C, it has been shown that all nucleic acid purines (Flavin and Graff, 1951a,b; Heinrich *et al.*, 1953) and all free nucleotides (Heinrich *et al.*, 1953) can be derived from dietary guanine only. Formate and glycine are not incorporated into purines. Further, *T. pyriformis* does not have the ability to produce hypoxanthine by ring closure at position 2 or guanine by ring closure at position 8. The basis for these conclusions is the inactivity of formylaminoimidazole carboxamide and 2,6-diamino—4-hydroxy-5-formylaminopyrimidine in replacing or sparing the requirement for guanine (Kidder and Dewey, 1961).

The pyrimidine requirement of *T. pyriformis* can be satisfied by uracil, uridine, uridine monophosphate (UMP), cytidine, or cytidine monophosphate (CMP). Not effective in supplying the requirement are cytosine, orotic acid, thymine, thymidine, 5-methylcytosine, 5-methyldeoxycytidine, 5-methyluridine, thymidine monophosphate (TMP), and 5-methyldeoxyCMP (Kidder and Dewey, 1949a; Kidder *et al.*, 1950; Wykes and Prescott, 1968). There is an enhanced requirement for purines and pyrimidines at high temperatures (Allen, 1953), perhaps due to more rapid degradation of ribonucleic acid (Byfield and Scherbaum, 1966).

Tetrahymena species, *T. corlissi*, *T. limacis*, *T. patula*, *T. vorax*, and *T. paravorax*, have the same requirements as *T. pyriformis* (Erwin, 1960; Kessler, 1961; Holz *et al.*, 1961), but *T. setifera* is said to utilize, somewhat inefficiently, either cytosine or thymidine to fill the pyrimidine requirement (Holz *et al.*, 1962a). However, Holz (1964), in a review article, makes no point of this.

Purine and Pyrimidine Metabolism

A summary of the purine metabolism of *Tetrahymena* is given in Fig. 6.1 (reactions 1–30) (Hill and Chambers, 1967; Kidder, 1967). The presence of guanosine phosphorylase (reaction 1), inosine phosphorylase (reaction 2), guanine phosphoribosyltransferase (reaction 4), hypoxanthine phosphoribosyltransferase (reaction 5), and adenine phosphoribosyltransferase (reaction 6) can be demonstrated by enzyme assays and tracer experiments (Hill and Chambers, 1967); but adenosine phosphorylase (reaction 3) is apparently absent. Although initial assays for adenosine kinase (reaction 12) were negative (Hill and Chambers, 1967), it is now known to be present (Hill *et al.*, 1970), as indicated by the phosphorylation of 6-methylthiopurine ribonucleoside, a substrate for this enzyme (Schnebli *et al.*, 1967), by intact cells. There is no evidence for kinases acting on guanine (reaction 10) and inosine (reaction 11).

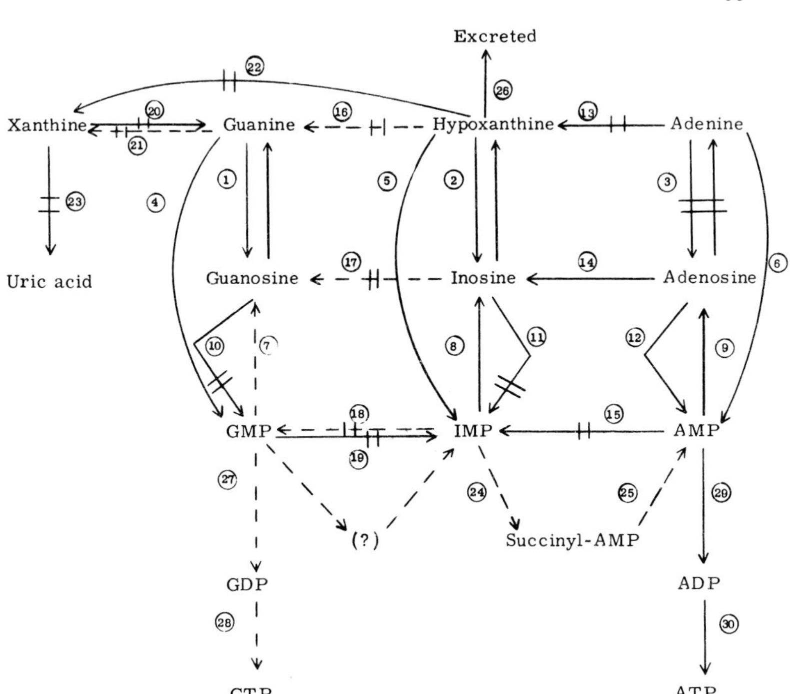

FIG. 6.1. Purine metabolism of *Tetrahymena*. Solid lines, reactions established by enzymatic assay; solid lines with crossbars, reactions for which the enzymatic assay is negative; dashed lines, reactions postulated to occur; dashed lines with crossbars are for reactions thought to be absent but for which no assay has been made. GMP, guanosine monophosphate. GDP, guanosine diphosphate; GTP, guanosine triphosphate IMP, inosine monophosphate; AMP, adenosine monophosphate; ADP, adenosine diphosphate; ATP, adenosine triphosphate.

Assays for xanthine oxidase (reactions 22 and 23) are negative (Hill and Chambers, 1967; Leboy *et al.*, 1964), and study of the metabolism of hypoxanthine under both aerobic and anaerobic conditions gives no evidence for production of xanthine or uric acid (Hill and Chambers, 1967). These results are in accord with the data of Wu and Hogg (1952), who initially investigated the nitrogen metabolism of *T. pyriformis*. However, Villela (1965) reports the existence of xanthine dehydrogenase in this organism, and Seaman (1963) presents meager spectrophotometric evidence for the presence of xanthine oxidase. But it appears that if such enzymes were physiologically significant, some xanthine or uric acid would have been found as a metabolite of hypoxanthine.

Guanase (reaction 20), adenase (reaction 13), and adenosine monophosphate (AMP) deaminase (reaction 15) are absent (Eichel, 1956; Conner and Linden 1970); but, characteristically, Seaman (1963) is able to find them. Deamination of adenosine (reaction 14) (Eichel, 1956); deoxyadenosine, and deoxyguanosine (Winicur and Roth, 1965; Conner and Linden, 1970) occurs. A nonspecific phosphatase that cleaves IMP is present (reaction 8) (Conner and McDonald, 1964); this enzyme probably acts on GMP (reaction 7) and AMP (reaction 9) (Conner and Linden, 1970). The inability of hypoxanthine or xanthine to replace guanine as a growth requirement (Kidder and Dewey, 1948) indicates the absence of reactions 16, 17, 18, and 21.

Since *Tetrahymena* can grow with guanine as the sole source of purines (Kidder and Dewey, 1948), some derivative of guanine must be converted to an adenine derivative. However, assays for GMP reductase (reaction 19) are negative (Hill and Chambers, 1967), which may indicate that removal of the amino group of guanine proceeds with a different enzyme, perhaps at the nucleoside di- or triphosphate level. Transformation of IMP to AMP perhaps proceeds though the intermediate succinyl-AMP (reactions 24 and 25). A kinase converting AMP to adenosine diphosphate (ADP) is present in cilia (reaction 29) (Culbertson, 1966) and most likely in other parts of the cell. The kinases for GMP, guanosine diphosphate (GDP), and ADP (reactions 27, 28, and 30) are evidently present.

Under adverse conditions for growth, *T. pyriformis* excretes hypoxanthine into the medium (reaction 26) (Leboy *et al.*, 1964; Cline and Conner, 1966), for it is the end product of purine catabolism. Hypoxanthine is also a principal excretory product of *Paramecium* (Soldo *et al.*, 1966).

The pyrimidine metabolism of *Tetrahymena* is summarized in Fig. 6.2 (reactions A–V). Cytosine is nutritionally inert (Kidder and Dewey, 1945; Wykes and Prescott 1968); thus it seems unlikely that it would be converted to cytidine (reaction A), to uracil (reaction L), or to CMP (reaction D), compounds which are readily utilized. Assays for cytosine deaminase (Eichel, 1957) and cytosine phosphoribosyltransferase are negative, but cytidine deaminase (reaction M) and uridine phosphorylase (reaction B) are present (Eichel, 1957; Hill and Chambers, 1967; Conner and Linden, 1970). Thymidine phosphorylase (reaction C) (Friedkin, 1953) and uracil phosphoribosyltransferase (reaction E) (Heinrikson and Goldwasser, 1964; Chambers and Brockman, 1965; Plunkett and Moner, 1970) are also active; but the rate of thymine incorporation into deoxyribonucleic acid is only 1/27 that for thymidine (Friedkin and Wood, 1956)—an indication that thymidine phosphorylase is not very active in the direction of the thymidine synthesis. Hill and Chambers (1967) and

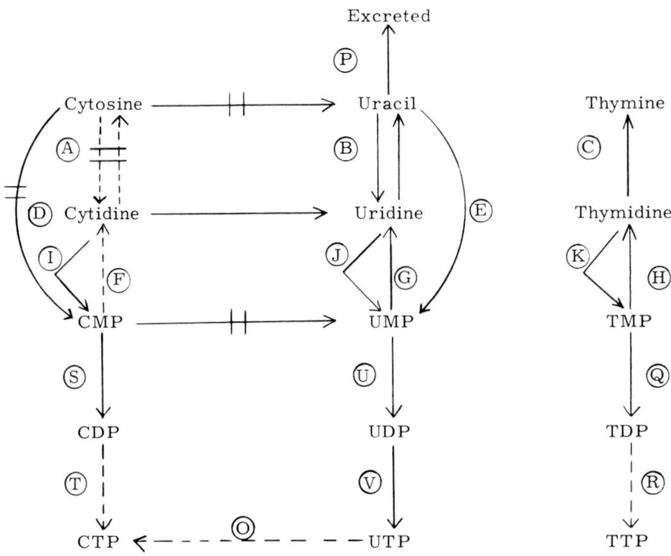

FIG. 6.2. Pyrimidine metabolism in *Tetrahymena*. Solid lines, reactions established by enzymatic assay; solid lines with crossbars, reactions for which enzymatic assay is negative; dashed lines, reactions postulated to occur; dashed lines with crossbars, reactions thought to be absent but for which no assay has been made. CMP, cytidine monophosphate; CDP, cytidine diphosphate; CTP, cytidine triphosphate; UMP, uridine monophosphate; UDP, uridine diphosphate; UTP, uridine triphosphate, TMP, thymidine monophosphate; TDP, thymidine diphosphate; TTP, thymidine triphosphate.

Ruffner and Anderson (1969) could not find activity for cytidine or uridine kinase (reactions I and J), but Heinrikson and Goldwasser (1964) reported weak activity in extracts of this organism. A brief report regarding some properties of a highly purified uridine-cytidine kinase has appeared (Plunkett and Moner, 1969). Shoup *et al.* (1966) found a thymidine kinase (reaction K) which is not inhibited by thymidine triphosphate (TTP) or any other pyrimidine deoxyribonucleotide; this property is in contrast to the enzyme in mammalian cells (Ives *et al.*, 1963). In fact, with the thymidine kinase of *T. pyriformis*, TTP and deoxycytidine triphosphate can substitute for adenosine triphosphate (ATP) as the donor of a phosphoryl group. Since cultures in logarithmic and stationary phases have thymidine kinase in equal quantities, the enzyme is thought to be constitutive (Shoup *et al.*, 1966). This may not be true, however, for it has been reported that individual cells must resynthesize the enzymes making

TTP with each cell cycle (Stone and Prescott, 1965). The activity of thymidylate kinase (reaction Q) correlates with deoxyribonucleic acid synthesis in *T. pyriformis* (Grieshaber and Duspiva, 1970). The activity is twice as high during logarithmic growth as in the stationary phase; and in synchronized cultures, the kinase increases with the start of the S phase. Thymidylate kinase activity also increases when the inhibition by amethopterin is overcome by adding thymidine and proteose–peptone.

A lag period is observed before a drop in absorbancy of CMP in cell extracts known to have cytidine deaminase activity (Hill and Chambers, 1967). Presumably, CMP is first dephosphorylated and then deaminated. If so, reaction N would not occur. Winicur and Roth (1965) and Conner and Linden (1970) reach a similar conclusion regarding the deamination of deoxyCMP and CMP. The nonspecific phosphatases (Conner and McDonald, 1964; Conner and Linden, 1970) are probably capable of removing the phosphate group of CMP and also that of UMP and TMP (reactions F, G, and H).

Labeled uracil is incorporated without dilution into the cytosine and thymine of nucleic acids (Heinrich *et al.*, 1957). Thus, some derivative of uridine must be converted to these derivatives, a transformation that likely proceeds, as it does in other cells, at the nucleoside triphosphate level (reaction O). No TMP synthetase, which converts deoxyUMP to TMP, can be detected in cell extracts (Wykes and Prescott, 1968). However, thymidine, 5-methyldeoxycytidine, 5-methyldeoxyCMP, and TMP partially reverse growth inhibition caused by lack of folic acid in the medium, suggesting that folic acid is involved in placing the methyl group in the 5-position of pyrimidines (Wykes and Prescott, 1968).

An enzyme that synthesizes uridine diphosphate (UDP; reaction U), cytidine diphosphate (CDP; reaction S), and deoxyCDP from the corresponding monophosphates has been purified 300-fold from *T. pyriformis* (Ruffner and Anderson, 1969). The substrate concentrations at half-maximal velocity are 1.5×10^{-3} *M* for UMP and 7.4×10^{-4} *M* for CMP. The enzyme is inhibited by CDP, UDP, ADP, and ATP. There apparently are two binding sites for UMP, but CMP may bind at only one of the two. The kinase acting to produce uridine triphosphate (UTP) (reaction V) can also be demonstrated enzymatically (Ruffner and Anderson, 1969).

Paper chromatographic separation of thymidine metabolities of *T. pyriformis* showed that TMP, thymidine diphosphate (TDP); (reaction Q), and TTP (reaction R) are produced (Jacobson and Prescott, 1964). Two unidentified metabolites are also present. A similar experiment with cytidine shows that CMP, CDP, and cytidine triphosphate (CTP; reaction T), UMP, UDP, and UTP are also formed.

An enzyme that uses uracil and ribose 5-phosphate to form pseudoUMP, a component of transfer ribonucleic acid, has been purified thirteenfold (Heinrikson and Goldwasser, 1964). The enzyme is distinct from uracil phosphoribosyltransferase. The optimum pH for the reaction is 8.5; and the Michaelis constants for uracil and ribose 5-phosphate are 7.1×10^{-5} *M* and 2.4×10^{-5} *M*, respectively. However, incorporation studies do not show that this enzyme is active in the formation of the pseudoUMP in transfer ribonucleic acid (Kusama *et al.*, 1966). The main point of these studies is that the ribosyl moiety of uridine labels the ribosyl portion of ribonucleic acid uridine and pseudouridine equally, demonstrating that neither free uracil nor diribosyluracil is an intermediate. The most obvious possibility and one that is consistent with the incorporation studies is that rearrangement of uridine occurs at the polynucleotide level, as in *Escherichia coli* (Johnson and Söll, 1970).

A kinetic study of sixty-fold purified deoxyuridine phosphorylase of *T. pyriformis* suggests that the enzyme binds uracil in two distinct ways—one catalytic, the other noncatalytic and inhibitory. The latter site is sensitive to urea (Lyon and Anderson, 1967).

Uracil is the end product of pyrimidine catabolism and is excreted into the medium (reaction P) (Cline and Conner, 1966).

The acid-soluble phosphate fraction of *T. pyriformis* contains AMP, GMP, UMP, CMP, UTP, ADP, and ATP (Kamiya, 1960). The ATP content is low during logarithmic-phase growth but increases during the deceleration phase (Cameron and Fisher, 1964). In cultures for which the oxygen tension is low and cell multiplication has almost ceased, the nucleoside triphosphate level is high (Scherbaum *et al.*, 1962). Aeration of the culture for 30 min increases the level even higher, but it falls after 2 hr of aeration. Apparently, a certain level of ATP is required for motility, for when the energy supply provided by respiration and glycolysis is cut off, intensity of movements decreases in parallel to the decreases in ATP content (Burnasheva and Efremenko, 1962; Burnasheva and Karusheva, 1967).

For cells being heat-treated for synchronous division, the nucleoside triphosphate concentration increases between the end of treatment and the first synchronized division (Plesner, 1958, 1961, 1964; Scherbaum *et al.*, 1962). On the other hand, there is no change between the content before heat treatment for 30 min at 29°C and 30 min after the end of treatment (Scherbaum *et al.*, 1962). The content drops during the heat period, but it rapidly recovers. Prolonged exposure (2 hr) at 34°C increases the level of nucleoside triphosphates from 7–13 to 16 nmoles/mg of protein. The level remains high for 24 hr.

Phosphatases

The acid phosphatases present in *Tetrahymena pyriformis* act on nucleotides and also on other phosphorylated compounds such as α-naphthyl phosphate and β-glycerophosphate. These enzymes, for the most part, are associated with cytoplasmic granules and food vacuoles and are necessary for digestion of food taken in through the buccal apparatus (Seaman, 1961; Allen, 1965; Bak and Elliott, 1963; Elliott and Bak, 1964; Loeh and Levy, 1969). The cytoplasmic granules containing acid phosphatases are identified as lysosomes (Müller *et al.*, 1966, 1968), for which the enzymes serve as a marker. Inert polystyrene–latex particles are ingested by the ciliate into food vacuoles, which rapidly acquire the acid phosphatase activity contributed by the lysosomes. Such action by the cell shows that there is no selection between nutritive and nonnutritive particles (Müller, 1965).

A portion of the acid phosphatase of lysosomes is latent but can be unmasked by adding Triton X-100 to the assay mixture (Lee, 1970a, 1971). The decrease in latency, produced by certain compounds, is interpreted as resulting from permeation of the lysosomes. Compounds effective in this regard are glycerol, dimethylsulfoxide, glucose, galactose, α-methyl glucose, mannitol, sorbitol, and inositol (Lee, 1970a), and some amino acids and some dipeptides (Lee, 1971). The latency is low during the logarithmic phase of growth, but it reaches a maximum just after the cell enters the stationary phase (Lee, 1970b).

Acid phosphatase activity is low in rapidly proliferating cells (Elliott and Hunter, 1951). On this point all investigators agree. Most agree that the activity is constant throughout the growth of a culture (Klamer and Fennell, 1963; Brightwell *et al.*, 1968; Lee, 1970b), except for the stage where the cells are rapidly dying (Klamer and Fennell, 1963). In contrast, Lazarus and Scherbaum (1968) found that it increases through the late logarithmic and stationary phases. One report states that acid phosphatase activity increases if the carbon source is omitted from the medium (Klamer and Fennell, 1963). This is disputed by other investigators (Lantos *et al.*, 1964) who found that it first decreases during starvation, although it later returns to the normal value. Acid phosphatase activity decreases with lowered pH of the medium and after heat treatment to induce synchronous division (Klamer and Fennell, 1963; Lasman, 1970).

The acid phosphatases of the cell supernatant have a pH optimum of 4.6 and are inhibited by (+)-tartarate and by cobalt, zinc, ferric, arsenate, and fluoride ions (Lazarus and Scherbaum, 1967c).

Acid phosphatase isozymes have been the subject of a series of genetic experiments (Allen *et al.*, 1963a,b, 1965; Allen, 1965, 1968). Each homozygous cell has one principal isozyme, and heterozygous cells have the two parental isozymes plus one to three new forms. As many as seventeen molecular variations may be present in some cells (see Chapter 7).

Concerning the specificity of a neutral, particle-bound phosphatase of *T. pyriformis* toward nucleotides, Connor and McDonald (1964) conclude that the enzyme is nonspecific with regard to the base, but is active on 3′-nucleotides and not on 5′-nucleotides. In a brief communication, Wu (1968) states that the 3′-nucleotidase has been partially purified.

Alkaline phosphatase activity is also present in *T. pyriformis* (Elliott and Hunter, 1951; Smith, 1961; Fennell and Degenhardt, 1957). This enzyme is located in the posterior region of the cell and acts on glycerophosphate, 5′-AMP, 3′-AMP, ATP, glucose 1-phosphate, and creatine phosphate (Fennell and Degenhardt, 1957). Alkaline phosphatase activity is at a minimum in 24-hr cultures and is absent in cells grown in a vitamin-enriched medium (Fennell and Marzke, 1954). This may be the explanation for the fact that no alkaline phosphatase could be detected by another group (Conner and McDonald, 1964).

A phosphodiesterase cleaving 3′,5′-cyclic AMP is present in *T. pyriformis* (Blum, 1970).

Effect of Purine Analogs

8-Azaguanine was the first purine analog shown to have an inhibitory effect on the growth of tumors (Kidder *et al.*, 1951). The basis for testing it against tumors was that it had earlier been found to inhibit the growth of *Tetrahymena pyriformis* (Kidder and Dewey, 1949b; Kidder *et al.*, 1949). This discovery opened a new field, and now purine and pyrimidine analogs are widely used as cancer chemotherapeutic agents. *Tetrahymena pyriformis*, because of its unusual requirements for preformed purines and pyrimidines, has continued to be a favorite organism for studying the effects of analogs, since clues to the mechanism of action of such compounds can be obtained by contrasting their effect on this protozoan with that on other organisms.

8-Azaguanine is incorporated into the ribonucleic acid of *T. pyriformis*, and very likely the formation of a faulty ribonucleic acid is responsible for its toxic effect (Heinrich *et al.*, 1952; Kidder *et al.*, 1952). This is the only purine analog inhibitory to the protozoan for which the site of action is strongly indicated. The effects of 8-azaguanine on *T. pyriformis* can be reversed by addition of adenine, hypoxanthine, or guanine (Flavin and Engelman, 1953; Dewey and Kidder, 1960; Holz *et al.*, 1962) or sterols (Dewey *et al.*, 1959) to the medium. 8-Azaguanosine is a less effective inhibitor than the free base (Dewey and Kidder, 1960). Another potent purine analog is 6-methylpurine (Dewey *et al.*, 1959). Its effect is reversed by adenine plus acetate, by sterols, or by phospholipids (Dewey *et al.*, 1960; Dewey and Kidder, 1962; Holz *et al.*, 1962b). Other purine analogs, less potent than 8-azaguanine and 6-methylpurine, which have been reported as inhibitors are 1-deazaguanine (Kidder and Dewey, 1957), 2-azaadenine, 8-azaadenine, purine, 6-chloropurine, 6-methoxypurine, 1-deaza-6-methylpurine (Dewey *et al.*, 1959), and 8-azaxanthine (Dewey and Kidder, 1960). A number of other purine analogs tested by these investigators are not strongly effective.

The pyrimidine analogs, 2-thio-4-aminopyrimidine and 5-hydroxyuracil are moderate inhibitors *T. pyriformis* growth (Kidder and Dewey, 1949a).

In an investigation for which the results have recently been published (Hill *et al.*, 1970), more than eighty purine and pyrimidine analogs have been tested for inhibition of growth of *T. pyriformis* in a proteose–peptone medium and in the chemically defined medium of Kidder and Dewey (1957). The results are shown in Table 6.1, but only those inhibitors that have a substantial effect are listed.

There is considerable information about some of the compounds tested. In cells possessing the *de novo* purine pathway, some purine analogs, after being converted to the nucleotide derivatives, act as pseudofeedback inhibitors of purine biosynthesis. 6-Methylthiopurine ribonucleoside, 6-thioguanine, and 6-mercaptopurine, as the nucleotides, are strong inhibitors of an early step of purine synthesis in various mammalian cell lines (LePage and Jones, 1961; Henderson, 1963; Brockman and Chumley, 1965; Bennett *et al.*, 1965). 2-Fluoroadenine (Bennett and Smithers, 1964a) and 2-fluoroadenosine (Bennett and Smithers, 1964b), presumably as ribonucleotides, moderately inhibit an early step of the *de novo* pathway in human epidermoid carcinoma cells. Direct assays of 5-phosphoribosyl–pyrophosphate amidotransferase of Adenocarcinoma 755 cells, which is the first enzyme in the purine biosynthetic pathway, reveal that the ribonucleotides of 6-methylthiopurine and 6-thioguanine are potent inhibitors; but the ribonucleotides of 2-fluoroadenine, 6-mercaptopurine, and 8-azaguanine have only a moderate effect (Hill and Bennett, 1969). The relatively weak inhibition by 6-mercaptopurine ribonucleotide on the isolated enzyme can be explained, for 6-methylthiopurine ribonucleotide is thought to be the common, active metabolite of both 6-mercaptopurine and 6-methylthiopurine ribonucleoside (Allan *et al.*, 1966).

Of the compounds that are known pseudofeedback inhibitors of the purine pathway, only 2-fluoroadenine, 2-fluoroadenosine, and 8-azaguanine are active against *T. pyriformis* (Hill *et al.*, 1970). This organism, lacking

TABLE 6.1

INHIBITION OF THE GROWTH OF *Tetrahymena pyriformis* BY PURINE AND PYRIMIDINE ANALOGS[a]

Analog	Concentration required for 50% inhibition (μM)	
	Proteose–peptone medium	Defined medium
5-Fluorodeoxycytidine	0.40	0.054
5-Fluorodeoxyuridine	0.46	0.046
5-Fluorouracil	1.37	1.6
2-Fluoroadenosine	8.0	1.9
2-Fluoroadenine	14	2.5
5-Fluorocytidine	16	1.5
5-Fluorouridine	16	1.0
9-Cycloheptyladenine	36	94
9-Isopropyladenine	69	195
9-Cyclopentyladenine	72	150
9-*n*-Pentyladenine	138	310
9-*n*-Propyladenine	158	290
9-*n*-Butyladenine	200	130
6-Benzylthiopurine	37	60
6-Methylpurine	48	3.8
2-Amino-6-chloro-8-aza-9-ethylpurine	60	138
N-(4,6-Dichloro-5-pyrimidinyl)–acetamide	80	100
6-Dimethylaminopurine	125	70
C-Adenosine[b]	210	24
Purine ribonucleoside	>440	54
8-Azaguanine	>220	59
1-Benzyl-6-mercaptopurine	160	440

[a] From Hill *et al.* (1970).

[b] The carbocyclic analog of adenosine in which a methylene group replaces the oxygen of the ribofuranose ring.

the *de novo* pathway, must possess other sites that are sensitive to their action. It is noteworthy that none of the three are potent pseudofeedback inhibitors. 2-Fluoroadenine may have an effect similar to that of 8-azaguanine, for it is converted to the nucleotide and subsequently incorporated into the nucleic acids of *T. pyriformis* (Hill *et al.*, 1970). Also, the phosphorylation of 6-methylthiopurine ribonucleoside (Hill *et al.*, 1970), a substrate for adenosine kinase (Schnebli *et al.*, 1967), implies that 2-fluoroadenosine, also a substrate for adenosine kinase, can be converted to the nucleotide by this organism. 2-Fluoroadenosine would, thus, appear to act in the same manner as 2-fluoroadenine. A conclusion is that feedback inhibition of purine biosynthesis by these three agents, in cells other than *T. pyriformis,* may not be responsible for growth inhibition.

The phosphoribosyltransferase converting 6-mercaptopurine and 6-thioguanine to the nucleotide derivatives is active in *T. pyriformis,* as is the kinase acting on 6-methylthiopurine ribonucleoside (Hill and Chambers, 1967; Hill *et al.*, 1970). Thus, lack of inhibition of *T. pyriformis* growth by these analogs is not due to lack of conversion to the nucleotides. These data imply that, in other organisms, the primary effect of these agents on growth inhibition is at the first step of purine biosynthesis.

In cells that possess the *de novo* purine pathway, 6-mercaptopurine ribonucleotide is also inhibitory at several other steps. The enzymes converting IMP to succinyl–AMP and succinyl–AMP to AMP are inhibited (Davidson, 1960; Hampton, 1962; Salser *et al.*, 1960) as is IMP dehydrogenase (Salser *et al.*, 1960; Atkinson *et al.*, 1963). These enzymes may not be present in *T. pyriformis,* IMP dehydrogenase probably is not, since it would have no reasonable function in an organism that requires guanine. If the enzymes are present, the degree of inhibition is not enough to inhibit growth.

In addition to its pseudofeedback effect, 6-thioguanine ribonucleotide inhibits ATP–GMP phosphotransferase (Miech *et al.*, 1967), an enzyme which should certainly be present in *T. pyriformis.* The lack of inhibition of growth of *T. pyriformis* by 6-thioguanine may mean that inhibition of this enzyme in other cells is not critical.

6-Benzylthiopurine inhibits several mammalian cell lines (Clarke *et al.*, 1958), perhaps by blocking the formation of 5-phosphoribosylpyrophosphate (Henderson and Khoo, 1965). The biosynthesis of this compound, a substrate for purine base phosphoribosyltransferases, is a possible site of action of 6-benzylthiopurine in *T. pyriformis.*

The 9-alkyl derivatives of adenine, for which there is no possibility of being converted to the nucleotide, have no feedback effect on the purine pathway of human epidermoid carcinoma cells (Bennett and Smithers, 1964a). The mechanism by which they inhibit growth is unknown.

Since most of the purine derivatives effective against *T. pyriformis* can be considered analogs of adenine, four of them were tested in the defined medium for reversal by this purine (Hill *et al.*, 1970). Large amounts of adenine completely reverse the inhibition by 2-fluoroadenine, an indication that this analog and adenine, or metabolites thereof, compete for enzyme sites in the cell. The inhibition by purine ribonucleoside is partially reversed, but that for 6-dimethylaminopurine and 9-cyclopentyladenine is not.

Of all the pyrimidine analogs tested, only those that are derivatives of 5-fluorouracil are strongly effective (Hill *et al.*, 1970). A product common to all of them, 5-fluorodeoxyUMP, is a potent inhibitor of TMP synthetase (Hartman and Heidelberger, 1961). In the cell, 5-fluorocytidine is presumably deaminated and cleaved to 5-fluorouracil, which can be converted to the nucleotide by the action of uracil phosphoribosyltransferase. Reduction of this nucleotide would give the active metabolite. Deamination of 5-fluorodeoxycytidine coupled with the action of a kinase would also produce 5-fluorodeoxyUMP.

Although 5-fluorodeoxyuridine is a potent inhibitor of growth in tubes inoculated with *T. pyriformis*, somewhat higher levels are required to block the growth of cells which have reached the logarithmic phase (Wykes and Prescott, 1968). 5-Fluorouracil and 5-fluorodeoxyuridine block cell division and oral morphogenesis in synchronized cells (Frankel, 1965; Cerroni and Zeuthen, 1962b).

Characterization of Nucleic Acids

Values reported for the total amount of ribonucleic acid per cell of *Tetrahymena pyriformis* are, in nanograms, 0.1–0.2 (Lane and Padilla, 1971; Cameron and Guile, 1965), 0.25 (6.0% of the dry weight) (Scherbaum, 1957), 0.40 (Cline and Conner, 1966), 1.28 (Iverson and Giese, 1957), and 0.4–0.5 (Akaboshi *et al.*, 1967). The actual value may vary since the ratios of ribonucleic acid to protein and to deoxyribonucleic acid increase with the growth rate (Leick, 1967, 1968) and the total amount of ribonucleic acid increases during the heat treatment used to produce synchronous division (Iverson and Giese, 1957; Lane and Padilla, 1971). It is unlikely, however, that there could be a fivefold increase; and, accordingly, the value of 1.28 ng per cell probably is in error. The base composition of the ribonucleic acid for several strains is summarized in Table 6.2. For each strain, the amount of guanine plus cytosine is low, and the purine/pyrimidine ratio is slightly greater than 1.

Leick and Plesner (1968), Kumar (1968), Weller *et al.* (1968), and Chi and Suyama (1970) have described the ribonucleic acids of the cytoplasmic ribosomal subunits of *T. pyriformis*. The 50 S (60 S) subunits contain 25 (26) and 5 S ribonucleic acids, and the 30 S (40 S) subunits contain 17 S (14 S) molecules. Polyribosomes with sedimentation values of 105, 130, and 170 S have ribonucleic acids sedimenting between 9 and 12 S. Of the total ribonucleic acid fraction, that from ribosomes has a low guanine plus cytosine content (36%) (Kumar, 1969).

Mitochondrial ribosomes (80 S) contain 21, 14, and 5 S ribonucleic acids (Chi and Suyama, 1970). The base compositions of these molecules are

TABLE 6.2

Base Composition of Ribonucleic and Deoxyribonucleic Acids of *Tetrahymena pyriformis*

Strain	Cytosine	Guanine	Uracil or thymine	Adenine	% Guanine + cytosine	Ratio of purine to pyrimidine	Reference
			Ribonucleic acid				
EU 6000	19.2	24.7	28.9	27.2	43.9	1.08	Wells (1962)
EU 6002	18.4	26.4	26.8	28.3	44.8	1.22	Wells (1962)
HSM	18.0	27.4	26.9	27.7	45.4	1.20	Wells (1962)
W	22.2	20.6	26.7	30.3	42.8	1.04	Cummins and Plaut (1962)
GL	19.3	19.8	30.0	30.9	39.1	1.03	Scherbaum (1957)
			Deoxyribonucleic acid				
EU 6000	14.5	14.5	35.8	35.2	29.0	0.99	Wells (1962)
EU 6002	14.5	9.4	39.0	37.1	23.9	0.84	Wells (1962)
HSM	14.9	11.9	36.6	36.6	26.8	0.94	Wells (1962)

% Guanine + cytosine	Species and strain of *Tetrahymena*	Reference
22	*T. patula* LFF	(Schildkraut *et al.*, 1962)
24	*T. rostrata*	(Sueoka, 1961)
26	*T. pyriformis* 1-A, 1-WH-52, 1-IL-12, 2-1, 5-1, 8-2, 6-1	(Sueoka, 1961)
28	*T. pyriformis* 3-1, 7-1, 9-1	(Sueoka, 1961)
30	*T. pyriformis* GL	(Sueoka, 1961)
30	*T. pyriformis* W	(Schildkraut *et al.*, 1962)
32	*T. pyriformis* E	(Sueoka, 1961)

different from those for cytoplasmic ribosomal ribonucleic acid. Hybridization studies show that 21 and 14 S molecules are complementary to 3.8 and 1.9% of the mitochondrial deoxyribonucleic acid, respectively, leading to the suggestion that mitochondrial deoxyribonucleic acid carries the genetic information for making mitochondrial ribosomal ribonucleic acids. There is no complementarity between mitochondrial deoxyribonucleic acid and cytoplasmic ribosomal ribonucleic acids.

Only 2% of the cellular ribonucleic acid is found in the macronucleus (Leick, 1969b), part of which is in peripheral nucleoli and part in small interchromatin particles (Swift *et al.*, 1964). Nuclear subribosomal ribonucleoprotein particles can be isolated by treatment of macronuclei with phenol–sodium dodecyl sulfate. This extraction yields two rapidly labeled species which sediment at 45 and 32 S and which represent precursors of ribosomal ribonucleic acid. In the process of conversion, the molecules are methylated. Extraction of nuclei after treatment with deoxycholate gives predominantly 25 S, but also a small amount of 17 S ribonucleic acid molecules (Leick, 1969b). By studying labeling kinetics, one can measure the pool size and turnover rates of macronuclear ribosomal ribonucleic acids (Leick and Andersen, 1970). The pool size of the 17 S molecules is 0.5% of the cytoplasmic 17 S ribosomal ribonucleic acid, and the turnover rate constant is 0.92 min^{-1}. Corresponding values for 25 S molecules are 1.6% and 0.29 min^{-1}. During growth most nuclear ribonucleic acid is found as a precursor of ribosomes, and very few, if any, finished ribosomes are present (Leick and Andersen, 1970).

Suyama (1967) has conducted experiments to determine the origins of the ribonucleic acids in the mitochondria of *T. pyriformis* (Swift *et al.*, 1964). Both transfer ribonucleic acid, with a sedimentation coefficient of 4 S, and ribonucleic acids characteristic of the mitochondria, with sedimentation coefficients of 14 and 18 S, are present. The latter are bound to submitochondrial particles. The transfer ribonucleic acid does not hybridize with mitochondrial deoxyribonucleic acid, but the 14 and 18 S molecules do. The experiments indicate that the transfer ribonucleic acid present in the mitochondria is not transcribed from mitochondrial deoxyribonucleic acid, but that the 14 and 18 S materials are made using the mitochondrial deoxyribonucleic acid as a code.

According to one report, the normal *T. pyriformis* cell contains 14 pg of deoxyribonucleic acid, representing 0.33% of the dry weight (Scherbaum, 1957; Scherbaum *et al.*, 1959). Other reports give a value of 10 (Woodard *et al.*, 1968), 11 (Flavell and Jones, 1970), 5 to 15 (Cameron and Guile, 1965; Lane and Padilla, 1971), and 30 pg per cell (Iverson and Giese, 1957; Akaboshi *et al.*, 1967). There is no decrease in deoxyribonucleic acid content during starvation (McDonald, 1958). As expected, rapidly growing cells accumulate double the normal amount before division (Walker and Mitchison, 1957; McDonald, 1958; Cleffman, 1968; Woodard *et al.*, 1968), but most nongrowing organisms also contain a double amount. In about 2% of the cells, there occurs an additional round of replication during one extended cell generation, resulting in a fourfold increase (Cleffman, 1968).

Immediately after division, different strains of *T. pyriformis* contain different amounts of deoxyribonucleic acid in their macronuclei (Cameron and Stone, 1964). However, the period of synthesis (the S phase) is of the same duration for all three strains. This points to a control mechanism regulating synthesis in a precise temporal manner independent of the amount of deoxyribonucleic acid present.

Cells subjected to heat treatment to induce synchrony accumulate extra amounts of deoxyribonucleic acid (Iverson and Giese, 1957; Scherbaum *et al.*, 1959; Lowy and Leick, 1969; Lane and Padilla, 1971), but the amount returns to normal following division. Treatment of cells with X rays produces a similar effect (Ducoff, 1956). In addition to quantitative changes, there are also changes in the quality of the *T. pyriformis* deoxyribonucleic acid during and after heat treatment to induce synchrony, as shown by different elution profiles on ECTEOLA-cellulose columns from cells at different stages of growth (Mita and Scherbaum, 1965). Further, the deoxyribonucleic acid from different stages of synchronous division varies in its susceptibility to acid hydrolysis (Holm, 1968a,b), with low amounts being more stable. Such variations are not seen in logarithmically growing cells.

The base ratios for deoxyribonucleic acids of three strains of *T. pyriformis* are given in Table 6.2. Also included in this table are percent values of guanine plus cytosine for various other species and strains (Wells, 1962; Sueoka, 1961; Schildkraut *et al.*, 1962). As for ribonucleic acid, the values for guanine plus cytosine are unusually low. 5-Methylcytosine is absent (Jones and Thompson, 1963). The molar ratios of the four bases do not change during heat treatment of cells (Scherbaum, 1957).

The density of nuclear deoxyribonucleic acid from *T. pyriformis*, as determined by CsCl density gradient centrifugation, is reported to be 1.691 (Sueoka, 1961), 1.685 (Parsons and Dickson, 1965), 1.692 (Flavell and Jones, 1970), and 1.688 (Suyama and Preer, 1965). Investigators using such gradients are cautioned about the possibility of preparations containing glycogen, for this contaminant may be mistaken for deoxyribonucleic acid (Brunk and Hanawalt, 1966, 1969).

In the macronuclei, deoxyribonucleic acid is present in small bodies 0.1–0.3 μ in diameter (Swift *et al.*, 1964; Gorovsky, 1965), and it comprises

15–19% of the dry weight of the structures (Ringertz *et al.*, 1967). Together, these bodies contain 20 (Gorovsky and Woodard, 1969; Nilsson, 1970), 40 (Cleffman, 1968), or 75 (Flavell and Jones, 1970) times as much deoxyribonucleic acid as micronuclei.

The nucleoli of *T. pyriformis* contain a small clump of deoxyribonucleic acid (Charret, 1969). The cistrons of these structures, presumably coding for ribosomal ribonucleic acid, continue to replicate past the S phase of the cell cycle.

Tetrahymena pyriformis was one of the first organisms known to contain mitochondrial deoxyribonucleic acid. Studies showing incorporation of thymidine into mitochondria implicated the presence of this material (Parsons, 1964). Mitochondrial deoxyribonucleic acid can be identified in the intact mitochondria by electron microscopy (Stone and Miller, 1965; Swift, 1965) and by microfluorophotometry (Yamada *et al.*, 1967); also it has been isolated by several workers (Stone, 1965; Suyama and Preer, 1965; Flavell and Jones, 1970). The molecular weight is 4×10^7 (Suyama, 1966; Brunk and Hanawalt, 1969) or 3×10^7 daltons (Flavell and Jones, 1970; Flavell and Follett, 1970); and the density, as determined by CsCl density gradient centrifugation, is 1.682–1.686 (Suyama and Preer, 1965; Flavell and Jones, 1970). Another value reported is 1.671 (Parsons, 1965). As shown by double-labeling, however, the value is actually identical to the density of the deoxyribonucleic acid from the whole cell (Brunk and Hanawalt, 1969). The purified material from mitochondria of this organism is doubled-stranded and linear, with a mean length of 17.6 μ. Each mitocondrion has a cluster of seven or eight strands of this size for a total amount of 3.7×10^{-16} gm (Suyama and Preer, 1965; Suyama and Miura, 1968; Flavell and Jones, 1970). Renaturation of mitochondrial deoxyribonucleic acid is rapid and also precise, for the melting temperature of the renaturation product (80.5°C) is the same as that for native material (Flavell and Jones, 1970). No cohesive ends are present, and the two complementary strands can be isolated by alkaline CsCl density gradients (Flavell and Jones, 1970). The presence of this genetic material in mitochondria provides strong support for the notion that mitochondria originate by self-division and that much of the genetic information for synthesis of mitochondrial proteins is carried in these bodies.

The hypothesis that mitochondria are self-duplicating is accompanied by a similar one concerning the basal bodies of cilia or kinetosomes. Randall (1959) and Randall and Disbrey (1965), from radioautographic studies with ^{3}H-thymidine and selective staining with acridine orange, report that deoxyribonucleic acid is present in these structures. They calculate that each basal body contains about 2×10^{-16} gm, and they suggest that this codes for enzymes catalyzing the production of proteins that assemble into the fibers of the cilium. However, another report states that preparations of these bodies contain less than 1% deoxyribonucleic acid, leading to the conclusion that this small amount is due to contamination (Hoffman, 1965). No deoxyribonucleic acid can be seen in the structures with the electron microscope (Swift *et al.*, 1964). Two investigations show that thymidine is not incorporated into the kinetosomes (Rampton, 1963; Pyne, 1968), so that self-replication of these bodies remains an open question.

Histones

Nearly half the nuclear proteins of *Tetrahymena pyriformis* are histones, i.e., basic polypeptides which bind to nucleic acids (Alfert and Goldstein, 1955; Lee and Scherbaum, 1966). Procedures are available for their isolation (Iwai *et al.*, 1965; Lee and Scherbaum, 1966; Piéri *et al.*, 1968). In purified preparations, the lysine content is high, with lysine-to-arginine ratios ranging from 3.4 to 5.3 (Lee and Scherbaum, 1966; VanWoude, 1964). There are six main types of histone, each with different amino acid sequences and molecular weights which correspond to the main types of mammalian histones (Iwai *et al.*, 1970).

Binding of the dye fast green indicates that cells in early exponential growth have histones in the nucleus and the cytoplasm, but those in the maximum stationary phase have them only in the nucleus (Scherbaum *et al.*, 1959). A similar experiment with the histone-binding dye bromphenol blue further demonstrates that histones can be present both in the nucleus and in the cytoplasm (Bolund and Ringertz, 1966). Histones extracted from macronuclei and micronuclei, show the same peaks on gel electrophoresis, but the fractions are present in different amounts in the two types of nuclei (Gorovsky, 1970).

Two classes of histone–deoxyribonucleate complexes from *T. pyriformis* can be characterized on the basis of their solubility in the presence of mono- and divalent cations (Bhagavan, 1967).

Changes in the histone/deoxyribonucleic acid ratio are associated with changes in the division rate of a cell population (Stone, 1969). Stationary-phase populations maintain a ratio of 2:1; but logarithmically growing cells exhibit a ratio of 1:1, which represents an "unmasking" of the deoxyribonucleic acid (Lee and Scherbaum, 1966). Not unexpectedly, the activity of nuclear ribonucleic acid polymerase is inversely proportional to the ratio; and the total amount of nuclear histone decreases prior to cell division in synchronized cultures (Lee and Byfield, 1969).

There are no substantial qualitative changes in the histone composition during the various culture growth stages (Lee and Scherbaum, 1966; Lee and Byfield, 1969). Only minor differences are observed in the electrophoretic patterns of histones from logarithmic and stationary-phase cells, as well as for cells subjected to heat treatment.

Ribonucleases and Deoxyribonucleases

The intracellular ribonucleases of *Tetrahymena pyriformis* are located in the lysosomes of the cell (Müller *et al.*, 1966, 1968). Early studies, at a time When it was thought that only one intracellular enzyme was present, showed that the optimum pH for activity is 5.5 and that an extract from lilac leaves and cobalt, nickel, and cupric ions are inhibitory (Roth, 1956, 1959). It is now known that there are at least three ribonuclease components (Lazarus and Scherbaum, 1967a,b). The three enzymes studied have been purified 400-fold (I), 215-fold (II), and 400-fold (III). Each hydrolyzes both purine and pyrimidine internucleotide bonds to produce 2′,3′-cyclic nucleotides that are subsequently converted to 3′-phosphates. Neither hydrolyzes deoxyribonucleic acid, *p*-nitrophenylphosphate, or bis-*p*-nitrophenylphosphate in the presence or absence of ethylenediaminetetraacetate (EDTA), but EDTA, citrate, and urea are stimulatory. The three components have different specificities. Polyadenylic and polyinosinic acids are hydrolyzed at rates in the order III > II > I, but the order for polycytidylic acid is I > II > III. Polyuridylic acid is degraded by all three at the same rate. A further investigation shows that two internal ribonucleases, one which produces 2′-AMP from polyadenylic acid and another which produces 3′-AMP, can be partially purified (Wu, 1968). 2′, 3′-Cyclic AMP is an intermediate in both cases. The latter component may be identical to III described above.

One report states that ribonuclease activity is low in rapidly growing cells but increases and reaches a maximum in the stationary phase (Lazarus and Scherbaum, 1968). However, another investigation shows that the activity decreases as cells leave the logarithmic phase (Lee, 1970b). A portion of the ribonuclease is latent, but can be activated by Triton X–100 (Lee, 1970b). The latency is maximal at the beginning of the stationary phase.

Although the activity can be protected from high temperatures by substrates, messenger ribonucleic acid is unstable at 37°C (Byfield and Scherbaum, 1966; Lazarus and Scherbaum, 1966). The degradation of this material by ribonuclease at elevated temperatures may be responsible for the synchronous division of cells which have been treated with heat

(Byfield and Scherbaum, 1967a; Byfield and Lee, 1970a; Christenssen, 1968). The decay of messenger ribonucleic acid is increased by all temperatures over the optimal growth temperature for each strain tested. Both nuclear and ribosomal ribonucleic acid are lost. Also, ribonucleic acid from cells at the end of heat treatment has a decreased ability to hybridize with deoxyribonucleic acid—a fact that further supports the concept of accelerated degradation of messenger ribonucleic acid at higher temperatures (Christenssen, 1970).

A rapid degradation of ribonucleic acid also occurs when cultures of *T. pyriformis* in the logarithmic phase of growth are suspended in an inorganic, buffered medium. Thirty percent of the total cellular ribonucleic acid is degraded in 3 hr (Cline and Conner, 1966). This degradation is partially under the control of the ionic content and the tonicity of the cellular environment (Cline, 1966), for the excretion of UV-absorbing materials is elevated by sodium ions in a medium isotonic to the culture fluid or by a hypertonic environment. Magnesium ions, glucose, and acetate prevent loss of the degradation products (Koroly and Conner, 1968).

In addition to having the internal ribonucleases, *T. pyriformis* and *T. rostrata* liberate a heat-stable, acid ribonuclease into the culture medium (Eichel *et al.*, 1963). This extracellular enzyme reportedly comprises 60–70% of the total for *T. pyriformis.*

Deoxyribonuclease activity (type II) is present in *T. pyriformis* (Eichel and Roth, 1953; Roth and Eichel, 1955; Haessler and Cunningham, 1957; Smith, 1961; Holm, 1966a,b; Müller *et al.*, 1966; Brightwell *et al.*, 1968). The activity is found in culture filtrates as well as in the cell, and there is no activation by divalent cations (Haessler and Cunningham, 1957). The pH optimum is between 5.0 and 5.5 (Holm, 1966b; Müller *et al.*, 1966). The activity is inhibited by magnesium and sodium ions and is inactivated by acidic polymers of the heparin type (Holm, 1966b) and by cation exchange resins (Holm, 1969b). X-Irradiation of cell homogenates results in appreciable destruction of the activity (Roth and Eichel, 1955). Until recently, there was no indication of more than one enzyme in the organism, but there is now preliminary evidence for the presence of three distinct acid deoxyribonucleases (Holm, 1969a).

Biosynthesis of Ribonucleic Acids

The nucleus of *Tetrahymena pyriformis* is the major site of synthesis of ribonucleic acid. The enzyme accomplishing the reaction, ribonucleic acid polymerase, transcribes the messages carried by deoxyribonucleic acid. Its activity can be demonstrated in purified macronuclei (Lee and Byfield,

1970). The reaction depends on the presence of deoxyribonucleic acid, magnesium ions, and all four ribonucleoside triphosphates. The pH optimum is 8.9. The activity is stimulated by KCl and is especially high when manganese ions and ammonium sulfate replace the magnesium ions and KCl. A potent inhibitor of bacterial ribonucleic acid polymerase, rifampicin, is without effect on the *in vitro* enzyme system and on the growth of the organism (Byfield *et al.*, 1970). The product is ribonucleic acid for which nucleotides have been incorporated nonterminally. Its composition resembles that of cellular deoxyribonucleic acid (Wells, 1962) but not that of ribosomal ribonucleic acid (Lyttleton, 1963) or whole-cell ribonucleic acid (Scherbaum, 1957), suggesting that it may function as a messenger (Lee and Byfield, 1970).

Soluble extracts of the cells in 50% glycerol contain an enzyme capable of catalyzing the incorporation of ATP and CTP into transfer ribonucleic acid (Fitt, 1966), but prior treatment of the transfer ribonucleic acid with snake venom phosphodiesterase greatly enhances its ability to prime CTP incorporation, suggesting that the enzyme accomplishes only terminal addition of CMP and AMP to ribonucleic acid fragments.

Studies on the synthesis of ribonucleic acid during the cell cycle of normal cells show that the rate increases to a maximum near the middle of the interfission period and drops thereafter (Cleffman, 1965). This information differs from an earlier report (Prescott, 1960a), which states that the rate increases near the end of the period.

Following deprivation of pyrimidines, protein synthesis and cell division can occur without ribonucleic acid synthesis for approximately two generations (Lederberg and Mazia, 1960; Cameron, 1963, 1965). The messenger ribonucleic acid must be quite stable. After replacement of the pyrimidine supply, synthesis of ribonucleic acids starts immediately, followed by protein and, later, deoxyribonucleic acid synthesis; and the cells proceed to divide synchonously (Cameron, 1963, 1965).

During the heat treatments used to produce synchronous division, synthesis of both ribo- and deoxyribonucleic acid occurs (Iverson and Giese, 1957; Scherbaum *et al.*, 1959); but synthesis of heterogeneous, high molecular weight ribonucleic acid is selectively inhibited (Yuyama and Zimmerman, 1969). It is reported that synthesis of all ribonucleic acid ceases near the end of the heating period (Moner, 1965, 1967; Bernstein and Zeuthen, 1966), and this is suggested as a mechanism for induction of synchrony. However, these results are probably due to an inadequate assay (Byfield and Lee, 1970a). During the time between divisions of synchronized cells in growth-supporting media, the rate of ribonucleic acid synthesis is constant (Byfield and Scherbaum, 1967a; Byfield and Lee,

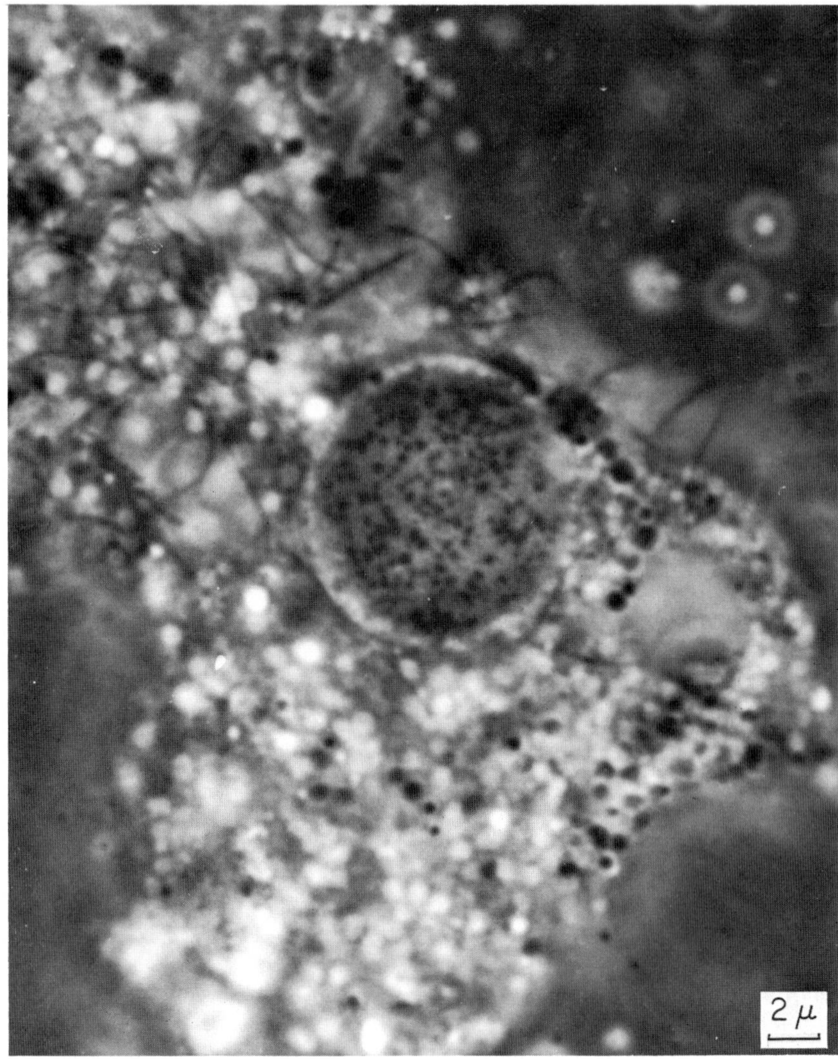

FIG. 6.3. Phase-contrast photograph showing nucleoli in the logarithmic phase of growth. Magnification: ×3500. (Courtesy of Dr. I. L. Cameron and Mr. Glenn Williams.)

1970a), although there is a reduction in the rate of incorporation of extracellular precursors. The constant rate is in contrast to the variable rate in normal cells. There is no net synthesis of nucleic acids during division (Lane and Padilla, 1971). The presence of amino acids stimulates synthesis of ribosomal ribonucleic acid but the absence of amino acids does not affect the rate of synthesis of division-related messenger ribonucleic acid (Byfield and Scherbaum, 1968a,b).

Ribonucleic acid transfer from central bodies of the nucleus to nucleoli can be detected (Elliott *et al.*, 1962). During the early logarithmic phase of growth, the nucleoli are separated; but late in logarithmic growth (Cameron and Guile, 1964, 1965), in the presence of actinomycin D (Satir and Dirksen, 1971), after heat shocks (Cameron *et al.*, 1966) or during starvation (Nilsson and Leick, 1970), they aggregate and fuse (Figs. 6.3–6.6). Such aggregation slows the production of ribosomes for protein synthesis. Addition of fresh medium to starving cells or to those in late logarithmic growth causes disaggregation, followed by increased ribonucleic acid synthesis and then by increased protein synthesis.

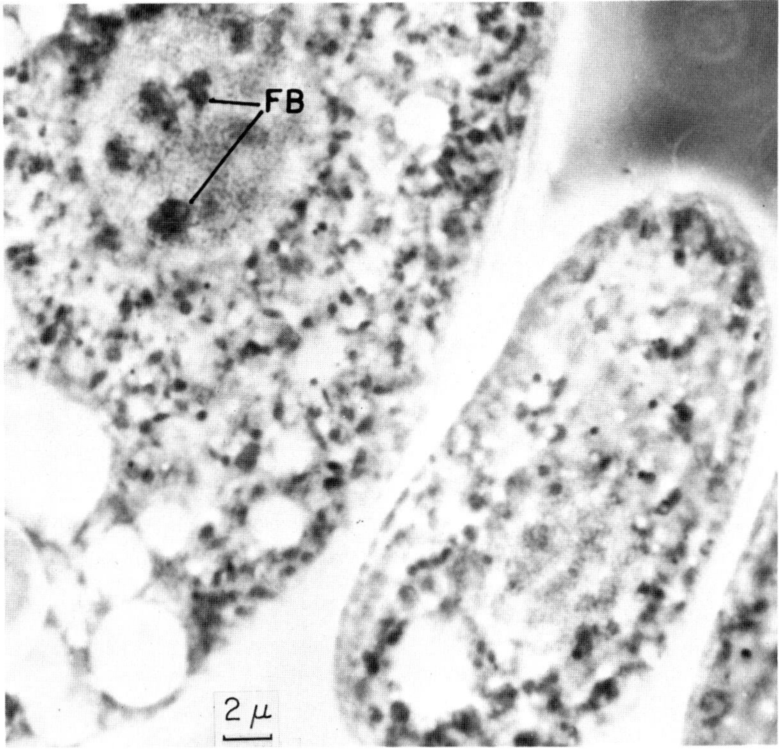

Fig. 6.4. Phase contrast photograph showing nucleoli in the stationary phase of growth. Fusion bodies (FB) are evident. Magnification: ×3500. (Courtesy of Dr. I. L. Cameron and Mr. Glenn Williams.)

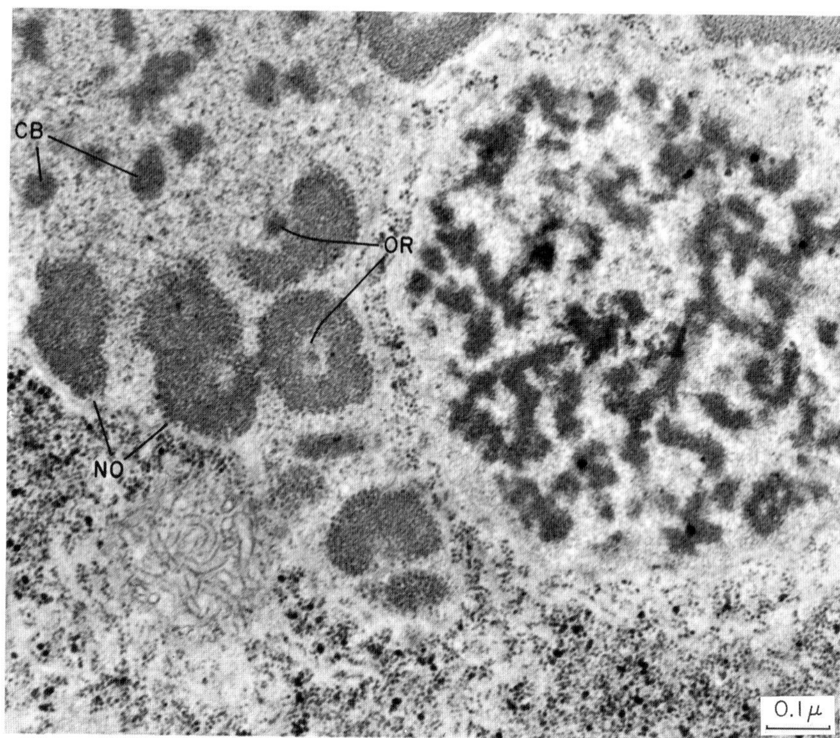

FIG. 6.5. Log-phase nucleoli (NO) showing nucleolar organizer (OR) and chromatin bodies (CB). The micronucleus is to the right. Magnification: ×125,000. (Courtesy of Dr. I. L. Cameron and Mr. Glenn Williams.)

Ribosomal ribonucleic acids are synthesized as 45 and 35 S (32 S) molecules in the macronucleus and later appear as 26 and 17 S molecules in the cytoplasm, where they are components of 60 and 45 S ribonucleoprotein particles (Leick, 1969b; Kumar, 1968, 1970). Although both species migrate rapidly from the nucleus, the 17 S molecules reach the cytoplasm first.

Ribonucleic acid synthesis in the cytoplasm can be detected radioautographically (Alfert and Das, 1959). Such synthesis probably occurs in the mitochondria, where ribonucleic acids with sedimentation constants of 14 and 18 S are made (Suyama and Eyer, 1968). Transfer ribonucleic acid is not made here, but the leucyl transfer ribonucleic acid and the leucyl transfer ribonucleic acid synthetase of mitochondria differ from those in the cytoplasm (Suyama and Eyer, 1967). The mitochondria contain a transfer ribonucleic acid fraction which can be labeled with ^{3}H-ATP but not with ^{3}H-UTP. Such information indicates that the labeling occurs at the terminal portions of the chains.

There are conflicting reports regarding ribonucleic acid synthesis by the micronucleus of sexually active strains. Two groups of investigators, using labeled precursors of ribonucleic acid, could find no incorporation (Alfert and Das, 1959; Gorovsky, 1965; Gorovsky and Woodard, 1969). However, the most recent report, by workers using tritiated precursors and the electron microscope, show incorporation into micronuclei which is removed by treatment with ribonuclease (Murti and Prescott, 1970). The incorporation coincides with the micronuclear S phase. Ribonucleic acid is mutually exchanged during conjugation of these cells (McDonald, 1966).

Actinomycin D blocks the transcription of ribonucleic acid from deoxyribonucleic acid (Hurwitz *et al.*, 1962), and this effect can be seen in *T. pyriformis* (Whitson and Padilla, 1964; Moner, 1965, 1967; Moner and Berger, 1966; Gavin and Frankel, 1969; Jauker, 1969). Actinomycin D can be used to show that ribonucleic acid synthesis is necessary for formation of oral structures and for the completion of cell division in cells undergoing synchronous division (Lazarus *et al.*, 1964; Whitson and Padilla, 1964; Whitson, 1965; Nachtwey and Dickinson, 1967; Gavin and Frankel,

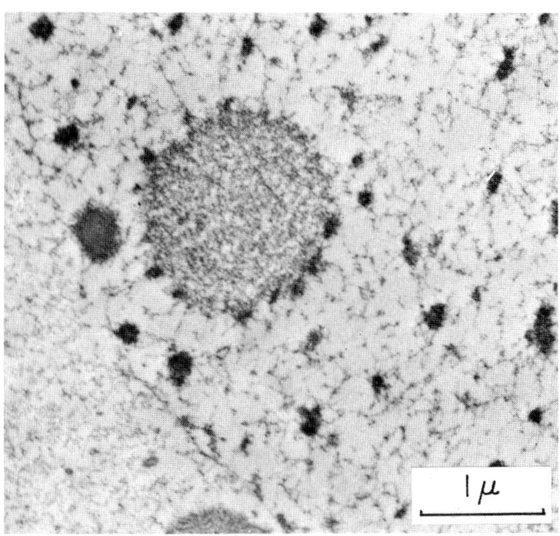

Fig. 6.6. Stationary-phase fused nucleoli. Magnification: ×16,000. (From Cameron and Guile, 1965.)

1969; Frankel, 1970); but in order to be effective, the drug must be added at a certain time before these events occur (Moner, 1965; Lazarus *et al.*, 1964; Mita, 1965). If actinomycin D is added 50 min prior to the first synchronous division, this division will not be affected; but the second one will be completely blocked (Mita, 1965; Lazarus *et al.*, 1964; Nachtwey and Dickinson, 1967; Dugaiczyk and Eiler, 1968). Cells synchronized by cold treatment are also sensitive to this agent (Moner and Berger, 1966; Moner, 1967). The synthesis of ribosomal ribonucleic acid is particularly sensitive to actinomycin D (Cristensson, 1968).

At the end of the first cell cycle after exposure of cells to a concentration of actinomycin D inhibiting 80% of ribonucleic acid synthesis, cells are larger and have about twice as much deoxyribonucleic acid as untreated ones (Jauker, 1969). The increase is due to an additional S phase. Culture growth recovers from the effect of actinomycin D by a process not involving selection of resistant cells, enzymatic inactivation, or alterations in the uptake of the agent. The additional deoxyribonucleic acid permits an increase in ribonucleic acid synthesis, which apparently allows the cells to adapt to actinomycin D (Jauker, 1969).

If actinomycin D is added at the time of inoculation of cultures, several effects on growth are noted (Satir, 1967). The lag phase, the logarithmic phase, and the generation time are increased; but the decelerating growth phase and the number of cells entering the stationary phase are decreased. Less total ribonucleic acid is made.

p-Fluorophenylalanine inhibits the incorporation of labeled precursors into ribonucleic acid (Rasmussen, 1968). This compound and canavanine, another amino acid analog, cause selective inhibition of the incorporation of orthophosphate into ribosomal ribonucleic acid but not into transfer ribonucleic acid (Leick, 1967, 1968, 1969a).

Biosynthesis of Deoxyribonucleic Acid

Studies on the formation of deoxyribonucleic acid by *Tetrahymena* often involve the use of labeled thymidine. After 48 hr of exposure to thymidine–methyl-^{3}H, 90% of the label can be found in deoxyribonucleic acid (Albach, 1967). Only the base thymine in the deoxyribonucleic acid is labeled; but small amounts of radioactivity are found in the acid-soluble fraction, in lipids, in ribonucleic acid, and in the cell residue. Wang *et al.* (1967) warn against the use of aged thymidine solutions, for the products of self-radiolysis are not incorporated into the deoxyribonucleic acid of *T. pyriformis* but are incorporated into other macromolecules in the cytoplasm. Perhaps this could account for some of the extraneous incorporation.

The soluble, thymidine derivative pool of *T. pyriformis* has been examined as a function of macronuclear deoxyribonucleic acid synthesis (Stone *et al.*, 1965). This pool, the existence of which has been confirmed in studies involving feeding of *T. pyriformis* with labeled bacteria (Calkins and Gunn, 1967), remains from one period of deoxyribonucleic acid synthesis (S phase) to another and does not turn over. Thymidine uptake by nuclei is almost entirely limited to the S phase; there is relatively little fixation during the presynthetic (G_1) or postsynthetic (G_2) phases.

The S phase in nonsynchronized cells lies largely in the first half of the interfission period (Prescott, 1960a,b; McDonald, 1962; Stone and Prescott, 1964a); but with decreased availability of nutrients, synthesis is initiated proportionately later in the cell division cycle (Cameron and Nachtwey, 1967). Also, the proportion of the cell cycle spent in deoxyribonucleic acid synthesis increases. One report states that synthesis can proceed throughout the interdivision period (Walker and Mitchison, 1957), but a definite S phase can be measured (Read and Buetow, 1971; Cameron and Nachtwey, 1967). Temperature has little effect on the length of the S phase, for between 19° and 20°C the generation time and G_2 double; but S remains the same (Cameron and Nachtwey, 1967). The micronuclei synthesize deoxyribonucleic acid in the late telophase and during a short time in the newly separated daughter cells (Stone and Prescott, 1964a). For these structures, the G_1 period is missing, the S period shortened, and the G_2 period greatly increased.

In cells treated to induce synchronous division, the rate of deoxyribonucleic acid synthesis is constant and asynchronous through the first 80% of the period leading to fission, after which time it drops off sharply (Cerroni and Zeuthen, 1962a; Hjelm and Zeuthen, 1967a,b; Padilla and Lane, 1969; Andersen *et al.*, 1969; Lane and Padilla, 1971). But a few minutes before division, synthesis is increased markedly and is evidently synchronized (Hjelm, 1968; Hjelm and Zeuthen, 1967a,b). After heat treatment, not all of the cells take up thymidine; and in the presence of inhibitors which stop deoxyribonucleic acid synthesis there is no blockade of division for the next two fissions, for there is no S phase (Cerroni and Zeuthen, 1962b; Hjelm and Zeuthen, 1967a,b; Byfield and Scherbaum, 1967b; Lowy and Leick, 1969; Jeffrey *et al.*, 1970). This is probably owing to the fact that the heat-treated cells have twice the deoxyribonucleic acid of untreated cells and do not need exogenously supplied thymidine (Scherbaum *et al.*, 1959).

A sizable portion of the deoxyribonucleic acid accumulated during heat treatment of *T. pyriformis* is extruded as small bodies by the cell (Scherbaum *et al.*, 1958; Sullivan and Rice, 1965; Hjelm and Zeuthen, 1967a,b). This phenomenon, known as genomic extrusion, is observed in 55% of the dividing cells following the first synchronized division, but in only 16% of the dividing cells in normal cultures (Scherbaum *et al.*, 1958). Extrusion occurs most often in cells under starvation conditions or in the early stationary growth phase (Nilsson, 1970). The mean volume of the extrusion bodies is 1.9 μ^3 for normal cells and 12.2 μ^3 for heat-treated cells. The extrusion body for *T. pyriformis* contains 2–25% of the macronuclear deoxyribonucleic acid; for *Tetrahymena patula* the amounts range from 4 to 16% (Dysart *et al.*, 1963). The extrusion of chromatin occurs in 96% of the vegetative divisions of *Tetrahymena limacis* (Dysart, 1960, 1963).

The deoxyribonucleic acid of cells at different stages of synchronous division has different elution profiles from ECTEOLA-cellulose columns (Mita and Scherbaum, 1965). These differences have not been further characterized, but they may be related to genomic extrusion. Perhaps similarly associated is the fact that the average amount of deoxyribonucleic acid per cell decreases just before division (Holm, 1966a).

Amino acid requirements for deoxyribonucleic acid synthesis and cell division have been studied by depriving cells of histidine and tryptophan at defined intervals in the interdivisional period. Deprivation after the start of deoxyribonucleic acid synthesis does not prevent its completion or the following cell division. Deprivation before the start of synthesis does not prevent synthesis but does prevent division and also limits the increase in deoxyribonucleic acid to 20% (Stone, 1963; Stone and Prescott, 1964b). It is concluded that some amino acid-dependent event, probably protein synthesis, occurs around the beginning of deoxyribonucleic acid synthesis. Further evidence for this is that cycloheximide and puromycin, inhibitors of protein synthesis, block deoxyribonucleic acid synthesis if they are added prior to the S phase (Byfield *et al.*, 1968; Byfield and Lee, 1970b). The event is not essential to the beginning of synthesis but is essential to its maintenance. It may actually be the synthesis of the enzymes producing TTP, for between consecutive S phases these enzymes disappear and must be resynthesized with each cycle (Stone and Prescott, 1965).

Total starvation of *T. pyriformis* causes cells to stall in the G_1 phase of the cell cycle (Cameron and Jeter, 1970). When nutrients are added to starved cells, they make ribonucleic acid and then protein before replicating their deoxyribonucleic acid and proceeding to cell division (Raff and Padilla, 1970). Some recent data, obtained from experiments with cycloheximide (Byfield and Lee, 1970b), conflict with the idea that protein synthesis is not required for the maintenance of deoxyribonucleic acid synthesis. The investigators envision replication-supporting proteins which are coded for by short-lived, temperature-sensitive templates.

To a small extent, tritiated thymidine is incorporated into the deoxyribonucleic acid of mitochondria (Stone and Miller, 1964; Stone and Prescott, 1964a; Parsons, 1965; Parsons and Rustad, 1968; Charret and André, 1968). The rate of incorporation can be increased by amino acid starvation or by treatment with actinomycin D or 5-fluorodeoxyuridine (Stone and Miller, 1964). An early report states that during the cell cycle of *T. pyriformis*, there is a periodicity of thymidine incorporation into mitochondrial deoxyribonucleic acid (Stone and Prescott, 1964a), cytoplasmic synthesis being correlated with the beginning of macronuclear synthesis. Later investigations showed that mitochondria incorporate labeled thymidine whether or not the nuclei are synthesizing and retain the radioactivity over a period of several divisions (Parsons and Rustad, 1968; Charret and André, 1968; Cameron, 1966). If the latter reports are true, a nuclear origin for mitochondrial deoxyribonucleic acid is excluded, and the hypothesis that mitochondria are produced by growth and division of preexisting structures is strengthened.

Some properties of a polymerase involved in the formation of *T. pyriformis* deoxyribonucleic acid are known (Pearlman and Westergaard, 1968, 1969; Westergaard and Pearlman, 1969). The enzyme requires magnesium ions and the deoxynucleoside triphosphates of adenine, guanine, cytosine, and thymine and is partially dependent on the addition of primer. The total activity remains constant during heat treatment and the subsequent synchronous divisions, but it can be increased sixfold by incubation of cultures with a combination of amethopterin, a folic acid analog, and uridine. The increase does not occur in the presence of inhibitors of protein or ribonucleic acid synthesis. Two distinct fractions of polymerase activity can be found on Sephadex chromatography of extracts of amethopterin-treated, thymidine-starved, or irradiated cells (Westergaard, 1969, 1970; Westergaard *et al.*, 1970). One fraction is increased thirty-five-fold over the normal level, leading to the suggestion that it may be the repair polymerase. Except for optimum salt concentration, the properties of the fractions are similar. The induced polymerase is, curiously, associated with mitochondria (Westergaard *et al.*, 1970). Induction of the enzyme is dependent on ribonucleic acid and protein synthesis but is not inhibited by chloramphenicol, implying that it is under nuclear, not mitochondrial, control. The activity is strongly inhibited by ethidium bromide, and is increased in cells grown in the presence of low amounts of this agent.

The combination of amethopterin and uridine inhibits the second synchronous division of heat-treated cells (Zeuthen, 1968), but the inhibition is overcome by thymidine. The same combination inhibits deoxyribonucleic acid synthesis strongly, but only after the initial stages of the synthetic period. An explanation for the effect of this combination is that amethopterin blocks the formation of TMP; and uridine interferes with the mechanism for transporting thymidine into the cell or into the macronucleus, thus causing thymidine starvation. Amethopterin is a potent inhibitor of TMP synthetase in L cells (Borsa and Whitmore, 1969a,b). Less well defined is the complete and selective, but transient, inhibition of deoxyribonucleic acid synthesis by uridine alone or thymidine alone (Villadsen and Zeuthen, 1970). Cells inhibited by the combination of amethopterin and uridine can be made to synthesize deoxyribonucleic acid synchronously by the addition of thymidine (Villadsen and Zeuthen, 1969, 1970; Zeuthen, 1970). The separate synchrony controls for division (by heat shocks) and for replication of deoxyribonucleic acid synthesis (by control of supply of thymidine) can be used to study the process of synchrony (Zeuthen, 1970). A conclusion from such studies is that synchronous division will occur only if a round of replication is separated from division by a time which equals or exceeds the time between replication and division in normal cells.

At the same time that deoxyribonucleic acid polymerase is active in the cell, other enzymes are producing histones, with which deoxyribonucleic acid is associated (Hardin *et al.*, 1967).

Vincristine, an alkaloid inhibitor of cell division, stops deoxyribonucleic acid synthesis in *T. pyriformis*; but it only moderately affects protein and ribonucleic acid synthesis (Slotnick *et al.*, 1966). The inhibition is prevented by riboflavin or flavin mononucleotide. A closely related compound, vinblastine, can be used to synchronize the division of *T. pyriformis* (Stone, 1968a,b). It was first thought that the drug had some effect on micronuclear division, but this was ruled out when amicronucleate strains were also found to be inhibited. The idea that deoxyribonucleic acid synthesis might be critical (Sedgely, 1968) has also been ruled out (Sedgely and Stone, 1969).

p-Fluorophenylalanine depresses thymidine incorporation if added 40 min before division (Rasmussen, 1968), but it evidently does not act directly on the synthesis of deoxyribonucleic acid.

When synchronized cultures of *T. pyriformis* are exposed to ultraviolet radiation of 254 nm for 30 to 60 sec, the rate of synthesis of deoxyribonucleic acid and the total amount synthesized are reduced (Harrington, 1960). Effects on the macronucleus and on the micronucleus are the same.

However, irradiation of individual cells under similar conditions leads to a threefold increase in the deoxyribonucleic acid made before the first division (Shepard, 1964, 1965). The excess occurs whether the cells are irradiated during the synthetic period or afterward, and it is extruded in large and numerous extrusion bodies. During the following divisions, the normal amount of deoxyribonucleic acid is synthesized.

Damage to the deoxyribonucleic acid of *T. pyriformis* induced by ultraviolet light or X rays is repaired by a process similar to the repair system present in bacteria (Christensson and Giese, 1956; Pettijohn and Hanawalt, 1964; Brunk and Hanawalt, 1967; Calkins, 1967; Whitson *et al.*, 1968). This repair process, which involves defect excision and subsequent resynthesis of the damaged section, occurs in the dark (see Chapter 9). To aid the repair process, the activity of deoxyribonucleic acid polymerase increases greatly after irradiation of the cells (Keiding and Westergaard, 1969; Westergaard and Pearlman, 1969). Another process for the repair of damage induced by ultraviolet light, photoreactivation, is indicated by a reduction in observed repair synthesis following exposure to visible light (Calkins, 1962).

REFERENCES

Akaboshi, M., Maeda, T., and Waki, A. (1967). *Biochim. Biophys. Acta* **138,** 596.

Albach, R. A. (1967). *J. Protozool.* **14,** 271.

Alfert, M., and Das, N. K. (1959). *Anat. Rec.* **134,** 523.

Alfert, M., and Goldstein, N. O. (1955). *J. Exp. Zool.* **130,** 403.

Allan, P. W., Schnebli, H. P., and Bennett, L. L., Jr. (1966). *Biochim. Biophys. Acta* **114,** 647.

Allen, M. B. (1953). *Bacteriol. Rev.* **17,** 125.

Allen, S. L. (1965). *Brookhaven Symp. Biol.* **18,** 27.

Allen, S. L. (1968). *Ann. N.Y. Acad. Sci.* **151,** 190.

Allen, S. L., Misch, M. S., and Morrison, B. M. (1963a). *J. Histochem. Cytochem.* **11,** 706.

Allen, S. L., Misch, M. S., and Morrison, B. M. (1963b). *Genetics* **48,** 1635.

Allen, S. L., Allen, J. M., and Licht, B. M. (1965). *J. Histochem. Cytochem.* **13,** 434.

Anderson, H. A., Brunk, C. F., and Zeuthen, E. (1969). *J. Protozool.* **16,** Suppl. 31.

Atkinson, M. R., Morton, R. K., and Murray, A. W. (1963). *Biochem. J.* **89,** 167.

Bak, I. J., and Elliott, A. M. (1963). *J. Protozool.* **10,** 21.

Bennett, L. L., Jr., and Smithers, D. (1964a). *Biochem. Pharmacol.* **13,** 1331.

Bennett, L. L., Jr., and Smithers, D. (1964b). Personal communication.

Bennett, L. L., Jr., Brockman, R. W., Schnebli, H. P., Chumley, S., Dixon, G. J., Schabel, F. M., Jr., Dulmadge, E. A., Skipper, H. E., Montgomery, J. A., and Thomas, H. J. (1965). *Nature (London)* **205,** 1276.

Bernstein, E., and Zeuthen, E. (1966). *C. R. Trav. Lab. Carlsberg* **35,** 501.

Bhagavan, N. V. (1967). *Proc. Nat. Acad. Sci. U. S.* **57,** 1726.

Blum, J. J. (1970). *Arch. Biochem. Biophys.* **137,** 65.

Bolund, L., and Ringertz, N. R. (1966). *Exp. Cell Res.* **44,** 606.

Borsa, J., and Whitmore, G. F. (1969a). *Mol. Pharmacol.* **5,** 303.
Borsa, J., and Whitmore, G. F. (1969b). *Mol. Pharmacol.* **5,** 318.
Brightwell, R., Lloyd, D., Turner, G., and Venables, S. E. (1968). *Biochem. J.* **109,** 42P.
Brockman, R. W., and Chumley, S. (1965). *Biochim. Biophys. Acta* **95,** 365.
Brunk, C. F., and Hanawalt, P. C. (1966). *Exp. Cell Res.* **42,** 406.
Brunk, C. F., and Hanawalt, P. C. (1967). *Science* **158,** 663.
Brunk, C. F., and Hanawalt, P. C. (1969). *Exp. Cell Res.* **54,** 143.
Burnasheva, S. A., and Efremenko, M. V. (1962). *Biokhimiya* **27,** 167.
Burnasheva, S. A., and Karusheva, T. P. (1967). *Biokhimiya* **32,** 270.
Byfield, J. E., and Lee, Y. C. (1970a). *J. Protozool.* **17,** 445.
Byfield, J. E., and Lee, Y. C. (1970b). *Exp. Cell Res.* **61,** 42.
Byfield, J. E., and Scherbaum, O. H. (1966). *J. Cell. Physiol.* **68,** 203.
Byfield, J. E., and Scherbaum, O. H. (1967a). *J. Cell. Physiol.* **70,** 265.
Byfield, J. E., and Scherbaum, O. H. (1967b). *Nature (London)* **216,** 1017.
Byfield, J. E., and Scherbaum, O. H., (1968a). *Nature (London)* **218,** 1271.
Byfield, J. E., and Scherbaum, O. H. (1968b). *Exp. Cell Res.* **49,** 202.
Byfield, J. E., Lee, Y. C., and Scherbaum, O. H. (1968). *J. Cell Biol.* **39,** 164A.
Byfield, J. E., Lee, Y. C., and Bennett, L. R. (1970). *Biochim. Biophys. Acta* **204,** 610.
Calkins, J. (1962). *Nature (London)* **196,** 686.
Calkins, J. (1967). *Int. J. Radiat. Biol.* **13,** 283.
Calkins, J., and Gunn, G. (1967). *J. Protozool.* **14,** 216.
Cameron, I. L. (1963). *J. Cell Biol.* **19,** 12A.
Cameron, I. L. (1965). *J. Cell Biol.* **25,** 9.
Cameron, I. L. (1966). *Nature (London)* **209,** 630.
Cameron, I. L., and Fisher, W. D. (1964). *Amer. Zool.* **4,** 408.
Cameron, I. L., and Guile, E. E. (1964). *J. Cell Biol.* **23,** 17A.
Cameron, I. L., and Guile, E. E., Jr. (1965). *J. Cell Biol.* **26,** 845.
Cameron, I. L., and Jeter, J. R. (1970). *J. Protozool.* **17,** 429.
Cameron, I. L., and Nachtwey, D. S. (1967). *Exp. Cell Res.* **46,** 385.
Cameron, I. L., and Stone, G. E. (1964). *Exp. Cell Res.* **36,** 510.
Cameron, I. L., Padilla, G. M., and Miller, O. L., Jr. (1966). *J. Protozool.* **13,** 336.
Cerroni, R. E., and Zeuthen, E. (1962a). *Exp. Cell Res.* **26,** 604.
Cerroni, R. E., and Zeuthen, E. (1962b). *C. R. Trav. Lab. Carlsberg* **32,** 499.
Chambers, P. C., and Brockman, R. W. (1965). *Bacteriol. Proc.* **65,** 94.
Charret, R. (1969). *Exp. Cell Res.* **54,** 353.
Charret, R., and André, J. (1968). *J. Cell Biol.* **39,** 369.
Chi, J. C. H., and Suyama, Y. (1970). *J. Mol. Biol.* **53,** 531.
Christensson, E. G. (1968). *Z. Biol.* **116,** 143.
Christensson, E. G. (1970). *J. Protozool.* **17,** 496.
Christensson, E., and Giese, A. C. (1956). *J. Gen. Physiol.* **39,** 513.
Clarke, D. A., Elion, G. B., Hitchings, G. H., and Stock, C. C. (1958). *Cancer Res.* **18,** 445.
Cleffman, G. (1965). *Z. Zellforsch. Mikrosk. Anat.* **67,** 343.
Cleffman, G. (1968). *Exp. Cell Res.* **50,** 193.
Cline, S. G. (1966). *J. Cell. Physiol.* **68,** 157.
Cline, S. G., and Conner, R. L. (1966). *J. Cell. Physiol.* **68,** 149.
Conner, R. L., and Linden, C. (1970). *J. Protozool.* **17,** 659.
Conner, R. L., and McDonald, L. A. (1964). *J. Cell. Comp. Physiol.* **64,** 257.
Culbertson, J. R. (1966). *Science* **153,** 1390.

Cummins, J. E., and Plaut, W. (1962). *Biochim. Biophys. Acta* **55,** 418.
Davidson, J. D. (1960). *Cancer Res.* **20,** 225.
Dewey, V. C., and Kidder, G. W. (1960). *Z. Allg. Mikrobiol.* **1,** 1.
Dewey, V. C., and Kidder, G. W. (1962). *Biochem. Pharmacol.* **11,** 53.
Dewey, V. C., Kidder, G. W., and Markees, D. G. (1959). *Proc. Soc. Exp. Biol. Med.* **102,** 306.
Dewey, V. C., Heinrich, M. R., Markees, D. G., and Kidder, G. W. (1960). *Biochem. Pharmacol.* **3,** 173.
Ducoff, H. S. (1956). *Exp. Cell Res.* **11,** 218.
Dugaiczyk, A., and Eiler, J. J. (1968). *Curr. Mod. Biol.* **1,** 333.
Dysart, M. P. (1960). *J. Protozool.* **7,** Suppl. 10.
Dysart, M. P. (1963). *J. Protozool.* **10,** Suppl. 8.
Dysart, M. P., Corliss, J. O., and de la Torre, L. (1963). *J. Protozool.* **10,** Suppl. 8.
Eichel, H. J. (1956). *J. Biol. Chem.* **220,** 209.
Eichel, H. J. (1957). *J. Protozool.* **4,** Suppl. 16.
Eichel, H. J., and Roth, J. S. (1953). *Biol. Bull.* **104,** 351.
Eichel, H. J., Conger, N., and Figueroa, E. (1963). *J. Protozool.* **10,** Suppl. 6.
Elliott, A. M., and Bak, I. J. (1964). *J. Cell Biol.* **20,** 113.
Elliott, A. M., and Hunter, R. L. (1951). *Biol. Bull.* **100,** 165.
Elliott, A. M., Kennedy, J. R., and Bak, I. J. (1962). *J. Cell Biol.* **12,** 312.
Erwin, J. (1960). Ph.D. thesis, Syracuse University.
Fennell, R. A., and Degenhardt, E. F. (1957). *J. Protozool.* **4,** 30.
Fennell, R. A., and Marzke, R. O. (1954). *J. Morphol.* **44,** 587.
Fitt, P. S. (1966). *J. Protozool.* **13,** 507.
Flavell, R. A., and Follett, E. A. C. (1970). *Biochem. J.* **119,** 61P.
Flavell, R. A., and Jones, I. G. (1970). *Biochem. J.* **116,** 155.
Flavin, M., and Engleman, M. (1953). *J. Biol. Chem.* **200,** 59.
Flavin, M., and Graff, S. (1951a). *J. Biol. Chem.* **191,** 55.
Flavin, M., and Graff, S. (1951b). *J. Biol. Chem.* **192,** 485.
Frankel, J. (1965). *J. Exp. Zool.* **159,** 113.
Frankel, J. (1970). *J. Exp. Zool.* **173,** 79.
Friedkin, M. (1953). *J. Cell. Comp. Physiol.* **41,** Suppl. **1,** 261.
Friedkin, M., and Wood, H., IV. (1956). *J. Biol. Chem.* **220,** 1956.
Gavin, R. H., and Frankel, J. (1969). *J. Cell. Physiol.* **74,** 123.
Gorovsky, M. A. (1965). *J. Cell Biol.* **27,** 37A.
Gorovsky, M. A. (1970). *J. Cell Biol.* **47,** 631.
Gorovsky, M. A., and Woodard, J. (1969). *J. Cell Biol.* **42,** 673.
Grieshaber, M. K., and Duspiva, F. (1970). *Z. Naturforsch.* **25,** 517.
Haessler, H. A., and Cunningham, L. (1957). *Exp. Cell Res.* **13,** 304.
Hampton, A. (1962). *J. Biol. Chem.* **237,** 529.
Hardin, J. A., Einem, G. E., and Lindsay, D. T. (1967). *J. Cell Biol.* **32,** 709.
Harrington, J. D. (1960). Ph.D. thesis, Catholic University of America Press, Washington, D. C.
Hartman, K.-U., and Heidelberger, C. (1961). *J. Biol. Chem.* **236,** 3006.
Heinrich, M. R., Dewey, V. C., Parks, R. E., and Kidder, G. W. (1952). *J. Biol. Chem.* **197,** 199.
Heinrich, M. R., Dewey, V. C., and Kidder, G. W. (1953). *J. Amer. Chem. Soc.* **75,** 1341.
Heinrich, M. R., Dewey, V. C., and Kidder, G. W. (1957). *Biochim. Biophys. Acta* **25,** 199.

Heinrikson, R. L., and Goldwasser, E. (1964). *J. Biol. Chem.* **239,** 1177.
Henderson, J. F. (1963). *Biochem. Pharmacol.* **12,** 551.
Henderson, J. F., and Khoo, M. K. Y. (1965). *J. Biol. Chem.* **240,** 3104.
Hill, D. L., and Bennett, L. L., Jr. (1969). *Biochemistry* **8,** 122.
Hill, D. L., and Chambers, P. (1967). *J. Cell. Physiol.* **69,** 321.
Hill, D. L., Straight, S., and Allan, P. W. (1970). *J. Protozool.* **17,** 619.
Hjelm, K. K. (1968). *J. Protozool.* **15,** Suppl. 36.
Hjelm, K. K., and Zeuthen, E. (1967a). *Exp. Cell Res.* **48,** 231.
Hjelm, K. K., and Zeuthen, E. (1967b). *C. R. Trav. Lab. Carlsberg* **36,** 127.
Hoffman, E. J. (1965). *J. Cell Biol.* **25,** 217.
Holm, B. (1966a). *Biochim. Biophys. Acta* **119,** 647.
Holm, B. (1966b). *Exp. Cell Res.* **41,** 12.
Holm, B. (1968a). *J. Protozool.* **15,** Suppl. 35.
Holm, B. (1968b). *Exp. Cell Res.* **53,** 18.
Holm, B. (1969a). *J. Protozool.* **16,** Suppl. 33.
Holm, B. (1969b). *J. Protozool.* **16,** 655.
Holz, G. G. (1964). *In* "Biochemistry and Physiology of Protozoa" (S. H. Hutner, ed.), Vol. 3, p. 199. Academic Press, New York.
Holz, G. G., Erwin, J. A., and Wagner, B. (1961). *J. Protozool.* **8,** 297.
Holz, G. G., Jr., Erwin, J., Wagner, B., and Rosenbaum, N. (1962a). *J. Protozool.* **9,** 359.
Holz, G. G., Rasmussen, L., and Zeuthen, E. (1962b). *C. R. Trav. Lab. Carlsberg* **33,** 289.
Hurwitz, J., Furth, J. J., Maloney, M., and Alexander, M. (1962). *Proc. Nat. Acad. Sci. U.S.* **48,** 1222.
Iverson, R. M., and Giese, A. C. (1957). *Exp. Cell Res.* **13,** 213.
Ives, D. H., Morse, P. A., and Potter, V. R. (1963). *J. Biol. Chem.* **238,** 1468.
Iwai, K., Shiomi, H., Ando, T., and Mita, T. (1965). *J. Biochem.* (*Tokyo*) **58,** 312.
Iwai, K., Hamana, K., and Yabuki, H. (1970). *J. Biochem.* (*Tokyo*) **68,** 597.
Jacobson, K. B., and Prescott, D. M. (1964). *Exp. Cell Res.* **36,** 561.
Jauker, F. (1969/1970). *Cytobiologie* **1,** 208.
Jeffrey, W. R., Stuart, K. D., and Frankel, J. (1970). *J. Cell Biol.* **46,** 533.
Johnson, L., and Söll, D. (1970). *Proc. Nat. Acad. Sci. U.S.* **67,** 943.
Jones, A. S., and Thompson, T. W. (1963). *J. Protozool.* **10,** 91.
Kamiya, T. (1960). *J. Biochem.* (*Tokyo*) **47,** 69.
Keiding, J., and Westergaard, O. (1969). *J. Protozool.* **16,** Suppl. 33.
Kessler, D. (1961). M.S. thesis, Syracuse University.
Kidder, G. W. (1967). *In* "Chemical Zoology" (M. Florkin and B. T. Scheer, eds.), Vol. 1, p. 93. Academic Press, New York.
Kidder, G. W., and Dewey, V. C. (1945). *Arch. Biochem.* **8,** 293.
Kidder, G. W., and Dewey, V. C. (1948). *Proc. Nat. Acad. Sci. U.S.* **34,** 566.
Kidder, G. W., and Dewey, V. C. (1949a). *J. Biol. Chem.* **178,** 383.
Kidder, G. W., and Dewey, V. C. (1949b). *J. Biol. Chem.* **179,** 181.
Kidder, G. W., and Dewey, V. C. (1951). *In* "Biochemistry and Physiology of Protozoa" (A. Lwoff, ed.), Vol. 1, p. 323. Academic Press, New York.
Kidder, G. W., and Dewey, V. C. (1957). *Arch. Biochem. Biophys.* **66,** 486.
Kidder, G. W., and Dewey, V. C. (1961). *Biochem. Biophys. Res. Commun.* **5,** 324.
Kidder, G. W., Dewey, V. C., Parks, R. E., and Woodside, G. L. (1949). *Science* **109,** 511.
Kidder, G. W., Dewey, V. C., Parks, R. E., and Heinrich, M. R. (1950). *Proc. Nat. Acad. Sci. U.S.* **36,** 431.

Kidder, G. W., Dewey, V. C., Parks, R. E., and Woodside, G. L. (1951). *Cancer Res.* **11,** 204.
Kidder, G. W., Dewey, V. C., and Parks, R. E., Jr. (1952). *J. Biol. Chem.* **197,** 193.
Klamer, B., and Fennell, R. A. (1963). *Exp. Cell Res.* **29,** 166.
Koroly, M. J., and Conner, R. L. (1968). *J. Protozool.* **15,** Suppl. 14.
Kumar, A. (1968). *J. Cell Biol.* **39,** 76A.
Kumar, A. (1969). *Biochim. Biophys. Acta* **186,** 326.
Kumar, A. (1970). *J. Cell Biol.* **45,** 623.
Kusama, K., Prescott, D. M., Froholm, L. O., and Cohn, W. E. (1966). *J. Biol. Chem.* **241,** 4086.
Lane, N. M., and Padilla, G. M. (1971). *J. Cell. Physiol.* **77,** 93.
Lantos, T., Müller, M., Törö, I., Druga, A., and Vargha, I. (1964). *Acta Biol.* (*Budapest*) **6,** Suppl. 22.
Lasman, M. (1970). *J. Protozool.* **17,** Suppl. 28.
Lazarus, L. H., and Scherbaum, O. H. (1966). *J. Cell. Physiol.* **68,** 95.
Lazarus, L. H., and Scherbaum, O. H. (1967a). *Nature* (*London*) **213,** 887.
Lazarus, L. H., and Scherbaum, O. H. (1967b). *Biochim. Biophys. Acta* **142,** 368.
Lazarus, L. H., and Scherbaum, O. H. (1967c). *Life Sci.* **6,** 2401.
Lazarus, L. H., and Scherbaum, O. H. (1968). *J. Cell Biol.* **36,** 415.
Lazarus, L. H., Levy, M. R., and Scherbaum, O. H. (1964). *Exp. Cell Res.* **36,** 672.
Leboy, P. S., Cline, S. G., and Conner, R. L. (1964). *J. Protozool.* **11,** 217.
Lederberg, S., and Mazia, D. (1960). *Exp. Cell Res.* **21,** 590.
Lee, D. (1970a). *Biochim. Biophys. Acta* **211,** 550.
Lee, D. (1970b). *J. Cell. Physiol.* **76,** 17.
Lee, D. (1971). *Biochim. Biophys. Acta* **225,** 108.
Lee, Y. C., and Byfield, J. E. (1969). *J. Cell Biol.* **43,** 78A.
Lee, Y. C., and Byfield, J. E. (1970). *Biochemistry* **9,** 3947.
Lee, Y. C., and Scherbaum, O. H. (1966). *Biochemistry* **5,** 2067.
Leick, V. (1967). *C. R. Trav. Lab. Carlsberg* **36,** 113.
Leick, V. (1968). *J. Protozool.* **15,** Suppl. 37.
Leick, V. (1969a). *Eur. J. Biochem.* **8,** 215.
Leick, V. (1969b). *Eur. J. Biochem.* **8,** 221.
Leick, V., and Andersen, S. B. (1970). *Eur. J. Biochem.* **14,** 460.
Leick, V., and Plesner, P. (1968). *Biochim. Biophys. Acta* **169,** 398.
LePage, G. A., and Jones, M. (1961). *Cancer Res.* **21,** 642.
Loeh, A. L., and Levy, M. R. (1969). *J. Cell Biol.* **43,** 82A.
Lowy, B. A., and Leick, V. (1969). *Exp. Cell Res.* **57,** 277.
Lyon, G. M., Jr., and Anderson, E. P. (1967). *Fed. Proc. Fed. Amer. Soc. Exp. Biol.* **26,** 843.
Lyttleton, J. W. (1963). *Exp. Cell Res.* **31,** 385.
McDonald, B. B. (1958). *Biol. Bull.* **114,** 71.
McDonald, B. B. (1962). *J. Cell Biol.* **13,** 193.
McDonald, B. B. (1966). *J. Protozool.* **13,** 277.
Miech, P. R., Parks, R. E., Jr., Anderson, J. H., Jr., and Sartorelli, A. C. (1967). *Biochem. Pharmacol.* **16,** 2222.
Mita, T. (1965). *Biochim. Biophys. Acta* **103,** 182.
Mita, T., and Scherbaum, O. H. (1965). *J. Biochem.* (*Tokyo*) **58,** 130.
Moner, J. G. (1965). *J. Protozool.* **12,** 505.
Moner, J. G. (1967). *Exp. Cell Res.* **45,** 618.

Moner, J. G., and Berger, R. O. (1966). *J. Cell. Physiol.* **67,** 217.
Müller, M. (1965). *J. Protozool.* **12,** 27.
Müller, M., Baudhuin P., and de Duve, C. (1966). *J. Cell. Physiol.* **68,** 165.
Müller, M., Hogg, J. F., and de Duve, C. (1968). *J. Biol. Chem.* **243,** 5385.
Murti, K. G., and Prescott, D. M. (1970). *J. Cell Biol.* **47,** 460.
Nachtwey, D. S., and Dickinson, W. J. (1967). *Exp. Cell Res.* **47,** 581.
Nilsson, J. R. (1970). *J. Protozool.* **17,** 539.
Nilsson, J. R., and Leick, V. (1970). *Exp. Cell Res.* **60,** 361.
Padilla, G. M., and Lane, N. M. (1969). *J. Cell Biol.* **43,** 100A.
Parsons, J. A. (1964). *J. Cell Biol.* **23,** 70A.
Parsons, J. A. (1965). *J. Cell Biol.* **25,** 641.
Parsons, J. A., and Dickson, R. C. (1965). *J. Cell Biol.* **27,** 77A.
Parsons, J. A., and Rustad, R. C. (1968). *J. Cell Biol.* **37,** 683.
Pearlman, R., and Westergaard, O. (1968). *J. Protozool.* **15,** Suppl. 37.
Pearlman, R., and Westergaard, O. (1969). *C. R. Trav. Lab. Carlsberg* **37,** 77.
Pettijohn, D. E., and Hanawalt, P. C. (1964). *J. Mol. Biol.* **9,** 395.
Piéri, M. J., Vaugien, C., and Camus, M. G. (1968). *Bull. Soc. Chim. Biol.* **50,** 1253.
Plesner, P. E. (1958). *Biochim. Biophys. Acta* **29,** 462.
Plesner, P. (1961). *Cold Spring Harbor Symp. Quant. Biol.* **26,** 159.
Plesner, P. (1964). *C. R. Trav. Lab. Carlsberg* **34,** 1.
Plunkett, W., and Moner, J. G. (1969). *Amer. Zool.* **9,** 1106.
Plunkett, W., and Moner, J. G. (1970). *J. Cell Biol.* **47,** 159A.
Prescott, D. M. (1960a). *Exp. Cell Res.* **19,** 228.
Prescott, D. M. (1960b). *Annu. Rev. Physiol.* **22,** 17.
Pyne, C. K. (1968). *C. R. Acad. Sci. Ser.* **D267,** 755.
Raff, E. M., and Padilla, G. M. (1970). *J. Protozool.* **17,** Suppl. 13.
Rampton, V. W. (1963). *Nature (London)* **195,** 195.
Randall, J. T. (1959). *J. Protozool.* **6,** Suppl. 30.
Randall, J. T., and Disbrey, C. (1965). *Proc. Roy. Soc. (London) Ser.* **B162,** 473.
Rasmussen, L. (1968). *J. Protozool.* **15,** Suppl. 37.
Read, J. A., and Buetow, D. E. (1971). *Exp. Cell Res.* **64,** 239.
Ringertz, N. R., Bolund, L., and Debault, L. E. (1967). *Exp. Cell Res.* **45,** 519.
Roth, J. S. (1956). *Exp. Cell Res.* **10,** 146.
Roth, J. S. (1959). *Ann. N.Y. Acad. Sci.* **81,** 611.
Roth, J. S., and Eichel, H. J. (1955). *Biol. Bull.* **108,** 308.
Ruffner, B. W., and Anderson, E. P. (1969). *J. Biol. Chem.* **244,** 5994.
Salser, J. S., Hutchinson, D. J., and Bass, M. E. (1960). *J. Biol. Chem.* **235,** 429.
Satir, B. (1967). *Exp. Cell Res.* **48,** 253.
Satir, B., and Dirksen, E. R. (1971). *J. Cell Biol.* **48,** 143.
Scherbaum, O. (1957). *Exp. Cell Res.* **13,** 24.
Scherbaum, O. H., Louderback, A. L., and Jahn, T. L. (1958). *Biol. Bull.* **115,** 269.
Scherbaum, O. H., Louderback, A., and Jahn, T. L. (1959). *Exp. Cell Res.* **18,** 150.
Scherbaum, O., Chou, S. C., Seraydarian, K. H., and Byfield, J. E. (1962). *Can. J. Microbiol.* **8,** 753.
Schildkraut, C. L., Mandel, M., Levisohn, S., Smith-Sonneborn, J. E., and Marmur, J. (1962). *Nature (London)* **196,** 795.
Schnebli, H. P., Hill, D. L., and Bennett, L. L., Jr. (1967). *J. Biol. Chem.* **242,** 1997.
Seaman, G. R. (1961). *J. Biophys. Biochem. Cytol.* **9,** 243.
Seaman, G. R. (1963). *J. Protozool.* **10,** 87.

Sedgley, N. N. (1968). *J. Cell Biol.* **39,** 121A.
Sedgley, N. N., and Stone, G. E. (1969). *Exp. Cell Res.* **56,** 174.
Shepard, D. C. (1964). *J. Cell Biol.* **23,** 86A.
Shepard, D. C. (1965). *Exp. Cell Res.* **38,** 570.
Shoup, G. D., Prescott, D. M., and Wykes, J. R. (1966). *J. Cell Biol.* **31,** 295.
Slotnick, I. J., Dougherty, M., and James, D. H., Jr. (1966). *Cancer Res.* **26,** 673.
Smith, I. B. (1961). *Diss. Abstr.* **22,** 1230.
Soldo, A. T., Godoy, G. A., and van Wagtendonk, W. J. (1966). *Fed. Proc. Fed. Amer. Soc. Exp. Biol.* **25,** 783.
Stone, G. E. (1963). *J. Cell Biol.* **19,** 68A.
Stone, G. E. (1965). *J. Exp. Zool.* **159,** 33.
Stone, G. E. (1968a). *J. Cell Biol.* **39,** 130A.
Stone, G. E. (1968b). *J. Cell Biol.* **39,** 559.
Stone, G. E. (1969). *J. Cell Biol.* **42,** 837.
Stone, G. E., and Miller, O. L., Jr. (1964). *J. Cell Biol.* **23,** 89A.
Stone, G. E., and Miller, O. L., Jr. (1965). *J. Exp. Zool.* **159,** 33.
Stone, G. E., and Prescott, D. M. (1964a). *J. Protozool.* **11,** Suppl. 24.
Stone, G. E., and Prescott, D. M. (1964b). *J. Cell Biol.* **21,** 275.
Stone, G. E., and Prescott, D. M. (1965). *Int. Soc. Cell Biol. Symp.* **4,** 95.
Stone, G. E., Miller, O. L., and Prescott, D. M. (1965). *J. Cell Biol.* **25,** 171.
Sueoka, N. (1961). *Cold Spring Harbor Symp. Quant. Biol.* **26,** 35.
Sullivan, W. D., and Rice, F. M. (1965). *Trans. Amer. Microsc. Soc.* **84,** 48.
Suyama, Y. (1966). *Biochemistry* **5,** 2214.
Suyama, Y. (1967). *Biochemistry* **6,** 2829.
Suyama, Y., and Eyer, J. (1967). *Biochem. Biophys. Res. Commun.* **28,** 746.
Suyama, Y., and Eyer, J. (1968). *J. Biol. Chem.* **243,** 320.
Suyama, Y., and Miura, K. (1968). *Proc. Nat. Acad. Sci. U.S.* **60,** 235.
Suyama, Y., and Preer, J. R., Jr. (1965). *Genetics* **52,** 1051.
Swift, H. (1965). *Amer. Natur.* **99,** 201.
Swift, H., Adams, B. J., and Larsen, K. (1964). *J. Roy. Microsc. Soc.* **83,** 161.
VanWoude, G. F., Jr. (1964). *Diss. Abstr.* **25,** 2215.
Villadsen, I. S., and Zeuthen, E. (1969). *J. Protozool.* **16,** Suppl. 31.
Villadsen, I. S., and Zeuthen, E. (1970). *Exp. Cell Res.* **61,** 302.
Villela, G. (1965). *Proc. Soc. Exp. Biol. Med.* **118,** 834.
Walker, P. M. B., and Mitchison, J. M. (1957). *Exp. Cell Res.* **13,** 167.
Wang, M., Zeuthen, E., and Evans, E. A. (1967). *Science* **157,** 436.
Weller, D. L., Raina, A., and Johnstone, D. B. (1968). *Biochim. Biophys. Acta* **157,** 558.
Wells, C. (1962). *Amer. Zool.* **2,** 457.
Westergaard, O. (1969). *J. Protozool.* **16,** Suppl. 33.
Westergaard, O. (1970). *Biochim. Biophys. Acta* **213,** 36.
Westergaard, O., and Pearlman, R. E. (1969). *Exp. Cell Res.* **54,** 309.
Westergaard, O., Marcker, K. A., and Keiding, J. (1970). *Nature (London)* **227,** 708.
Whitson, G. L. (1965). *J. Exp. Zool.* **160,** 207.
Whitson, G. L., and Padilla, G. M. (1964). *Exp. Cell Res.* **36,** 667.
Whitson, G. L., Francis, A. A., and Carrier, W. L. (1968). *Biochim. Biophys. Acta* **161,** 285.
Winicur, S., and Roth, J. S. (1965). *J. Protozool.* **12,** 166.
Woodard, J., Gorovsky, M., and Kaneshiro, E. (1968). *J. Cell Biol.* **39,** 182A.
Wu, C., and Hogg, J. F. (1952). *J. Biol. Chem.* **198,** 753.

Wu, J. J. (1968). *J. Protozool.* **15,** Suppl. 13.
Wykes, J. R., and Prescott, D. M. (1968). *J. Cell. Physiol.* **72,** 173.
Yamada, M., Takakusu, A., Yamamoto, K., and Iwata, S. (1967). *Arch. Histol. Jap.* **27,** 387.
Yuyama, S., and Zimmerman, M. (1969). *J. Cell Biol.* **43,** 162A.
Zeuthen, E. (1968). *Exp. Cell Res.* **50,** 37.
Zeuthen, E. (1970). *Exp. Cell Res.* **61,** 311.

CHAPTER 7

Biochemical Genetics

Nuclei

Like several other ciliates, *Tetrahymena* contains a polygenomic macronucleus. Possibly as a reflection of its polygenomic nature, the structure displays many separate bodies containing deoxyribonucleic acid (Sonneborn, 1947; Nilsson, 1970a,b,c). In cells undergoing morphogenesis, the nucleoplasm is clear with small, dense chromatin granules about 50 nm in diameter; but in interphase cells it is finely granular and filamentous (Roth and Minick, 1958; Nilsson, 1970b). On starvation, interphase cells develop into a third type, described as "opaque" (Nilsson, 1970b). Chromatin fibrils, 4–10 nm in width, can be determined by electron microscopy (Wolfe, 1967; Lee and Byfield, 1969; Abdel-Hameed, 1969; Nilsson, 1970c). Nucleoli, dense bodies 100–500 nm in diameter, contain deoxyribonucleic acid and are found in a peripheral layer of the macronucleus (Gorovsky, 1965; Charret, 1969).

For most strains of *Tetrahymena*, there is present a micronucleus, which, although much smaller, stains with more intensity than the macronucleus

(Alfert and Goldstein, 1955). The micronucleus of *T. pyriformis* contains a diploid set of five chromosomes, all of which can be distinguished during meiosis (Ray, 1956a,b). Cells in the process of losing the micronucleus, called "semiamicronucleate," are known (Elliott *et al.*, 1964). The complete loss usually leads to death during fission or conjugation. For *T. pyriformis*, the chief form of senescence, a decline in viability after conjugation, is due to the loss of the micronucleus by micronucleate strains (Allen *et al.*, 1967). However, those that survive are stable, for among the strains having no micronucleus are those that have been cultured for long periods in the laboratory (Nanney, 1957, 1959a; Wells, 1961; Allen *et al.*, 1967). Whatever intracellular functions the micronucleus has may be taken over by some part of the amicronucleate cell (Wells, 1961).

Most strains of *T. pyriformis* which contain micronuclei are sexually active; that is, they undergo conjugation, the normal sexual process. Sometimes amicronucleate cells can acquire a micronucleus by conjugation, but this is a very rare occurrence (Wells, 1961). Since cells lacking a micronucleus normally do not participate in conjugation, this is likely to be a primary function of the structure (Elliott and Hayes, 1953).

Conjugation

The temporal sequence of nuclear events that take place during conjugation are shown in Fig. 7.1 (Elliott and Nanney, 1952; Nanney, 1953; Elliott and Hayes, 1953; Allen, 1967b). After the cells are attached, the micronucleus in each conjugant undergoes meiosis. Of the four meiotic products in each cell, three disintegrate; but the fourth undergoes a mitotic division to give rise to two pronuclei—one migratory (♂) and one stationary (♀). The migratory pronuclei are reciprocally exchanged, and each stationary nucleus fuses with a migratory pronucleus. The derived diploid nucleus divides twice mitotically; and the two anterior products enlarge to form macronuclear anlagen, the precursors of macronuclei. Meanwhile, the old macronucleus disintegrates. One of the posterior nuclei becomes the new micronucleus, but the other disintegrates after separation of the cells. The micronucleus divides mitotically for the first cell division, in which each daughter cell (karyonide) receives one of the new macronuclei. Vegetative divisions occur until the next conjugation. The deoxyribonucleic acid of the degenerating macronucleus is not reutilized unless the cells are under starvation conditions (McDonald, 1959).

During conjugation, tubules which may allow exchange of cytoplasmic materials exist in the area of contact of the two cells (Elliott and Tremor, 1958). Ribonucleic acid and protein are transferred during conjugation

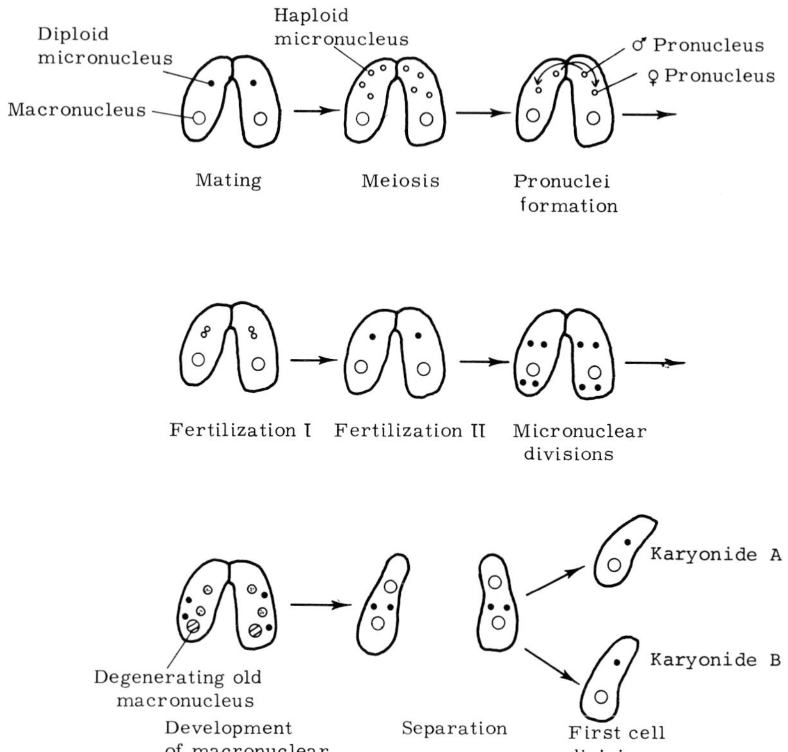

FIG. 7.1. Nuclear events in conjugation of *Tetrahymena pyriformis*. (Adapted from Allen, 1967c, and from Nanney, 1968.)

(McDonald, 1966; Schooley, 1958). Also, the infraciliature and the buccal apparatus dedifferentiate; they are redifferentiated shortly before the cells separate (Elliott and Tremor, 1958). There are no obvious, major changes in metabolism during conjugation; but there are fewer mitochondria. Those present are located near the cell wall (Elliott, 1959).

A number of factors influence the mating process. Starvation, easily accomplished by washing the cells in distilled water, is essential; but conjugation is prevented either by anoxia or by agitation (Elliott and Hayes, 1953). High levels of the amino acids or the vitamins commonly added to the medium reduce conjugation, but no specific amino acid or vitamin inhibits (Elliott and Hayes, 1953; Ducoff, 1956). Temperature, light, and age of the cells have little effect (Elliott and Hayes, 1953); and addition of either acetate, glucose, or purines and pyrimidines to the medium is not deleterious (Ducoff, 1956).

Following conjugation, the cells of some syngens (varieties) of *T. pyriformis* go through a period of immaturity during which they cannot mate (Nanney and Caughey, 1953; Gruchy, 1955; Orias, 1963); other syngens do not have this property (Gruchy, 1955; Elliott and Kennedy, 1962; Elliott *et al.*, 1962; Outka, 1961). Once maturity is achieved, the cell must mate again or the ability to mate may be lost, causing the cell to become senescent. There is recent information indicating that a heat-stable, low molecular weight substance is present in cultures activated for mating and that this material can induce mating in unreactive mixtures of cells of two mating types (Phillips, 1971).

Parthenogenic conjugation is reported to occur in crosses of one haploid clone of *T. pyriformis* previously treated with X-rays (Clark and Elliott, 1956; Elliott and Clark, 1956b). This clone fails to produce a migratory pronucleus, so that there is a one-way transfer of a pronucleus from diploid cells to haploid cells in crosses. Crosses of haploids give rise to amicronucleate cells.

Genomic Exclusion

An abnormal form of conjugation, occurring in syngen 1 of *Tetrahymena pyriformis,* is genomic exclusion (Allen, 1963; Nanney, 1963a; Nanney and Nagel, 1964). This phenomenon is revealed by the appearance of distorted genetic ratios from crosses of inbred strains. At present, true autogamy, conjugation within a specific mating type, has not been identified in *T. pyriformis,* but it is reported to occur in *Tetrahymena rostrata* (Corliss, 1952). Genomic exclusion, as for autogamy, results in complete homozygosity of the exconjugants, with the genetic contribution from one parent being lost. Thus, genomic exclusion provides a rapid means of producing homozygous, diploid lines (Allen, 1967a).

Genomic exclusion occurs in the mating of a cell having a defective micronucleus with a normal cell (Allen *et al.*, 1967). To produce homozygosity two rounds of mating are needed. In the first round, the defective cell receives a haploid pronucleus from the normal cell. The resulting haploid pronucleus in each cell undergoes duplication, and the products fuse to form a homozygous, diploid nucleus. The cells then separate. At this time the cells are heterokaryons, with a micronucleus of different genotype from that of the macronucleus; but only the macronuclear genes are expressed. If the exconjugants are reunited for a normal round of conjugation, the old macronucleus is lost; and the cells are homozygous.

Cells in the process of losing the micronucleus may participate in genomic exclusion. Since amicronucleate clones are ordinarily sexually dead, it may be that genomic exclusion provides the cell with a chance to replace the defective micronucleus and restore sexuality (Allen, 1967b; Allen *et al.*, 1967).

As stated above, genomic exclusion occurs in inbred strains. The process of inbreeding leads to various types of inbreeding degeneration in syngen 1 (Nanney, 1957; Nanney and Nagel, 1964). Death at conjugation and failure to complete meiosis rise in frequency for a few generations; but, later, survival improves until no lethality is noted. The degeneration may arise from the accumulation of homozygous, defective genes.

Mutants

Only a few mutants of *Tetrahymena pyriformis* are known. The induction of mutations by chemicals is very difficult, for the organism has a complex nutritional requirement and there is no easy method for screening and isolating mutants. In an experiment designed to detect differences in the nutritional requirement, Elliott and Hayes (1954) subjected a large number of clones to a variety of mutagenic agents but found no mutants.

Some clones have been described as either serine or pyridoxine mutants. Six of 2500 clones examined reportedly can grow without serine, and 41 of the 2500 ostensibly survive without a source of vitamin B_6 (Elliott and Hayes, 1955a; Elliott and Clark, 1956a). In each case, other amino acids and other vitamins are required. The wild-type requirers are said to be heterozygotes and the nonrequiring mutants, homozygous recessives (Elliott and Clark, 1958a,b), which is a very unusual situation. Further, at least one serine-requiring strain synthesizes serine from glucose (Elliott, 1959). The obvious question is—If the strain is able to make serine endogenously, why does it require exogenous serine?

Two recessive lethal mutations are known (Orias, 1957, 1960). The lethal conditions are controlled by unlinked, recessive genes *tt* (tiny) and *ff* (fat). Both are lethal when homozygous. A morphological mutant, which is multinucleate and bizarre-shaped, has been reported (Allen, 1958). The trait disappears at conjugation.

Mating Types

A number of sexually active strains of *Tetrahymena pyriformis* have been found in nature. More than fifty mating types, distributed among

12 syngens, are known (Elliott and Gruchy, 1952; Elliott and Hayes, 1955b; Gruchy, 1955; Elliott *et al.*, 1962, 1964). Their geographical distribution around the world is summarized in a recent review (Elliott, 1970). Since mating cannot take place between cells of different syngens, one might consider each syngen a separate species. However, the inclusive name *T. pyriformis* is not likely to be changed, as it is well established in the literature.

The mating-type inheritance of syngens 2 and 8 is under direct genic control (Hurst, 1957; Orias, 1958); that is, the genotype directly determines the mating properties of a cell. For syngen 8, there are three alleles that can be arrayed in the order of dominance (Orias, 1958). However, the mating-type alleles for syngen 8 "mutate" at a phenomenal rate of more than 4% per generation (Orias, 1963). Conjugation with syngen 5 reveals nuclear aberrations and yields no viable offspring, so that its breeding system remains unknown (Ray, 1956a; Ray and Elliott, 1954).

The mating-type inheritance of syngens 1, 6, 7, and 9 is not under direct genic control (Nanney and Caughey, 1953; Nanney *et al.*, 1955; Nanney, 1956, 1959b, 1968; Elliott and Clark, 1957, 1958b; Phillips, 1969). The genotype determines only an array of potentialities, the expression of which is governed by other factors. Most of the investigations on mating-type inheritance have been made with syngen 1. The mating-type expression in this syngen is described as "karyonidal." That is, there is heterogeneity within a particular karyonide (Nanney and Caughey, 1955). Karyonides are sexually immature, and individual mating types can be determined only after approximately eighty divisions (Nanney, 1958). However, mating-type potentialities are usually established at the first division following conjugation, making the expression of mating type independent of the process of maturation (Nanney, 1960). Segregation of the potentialities at this division is probably due to the segregation of the new macronuclei. Thus mating-type potentialities would be under macronuclear control.

The actual expression of mating types can be explained by the hypothesis that the polygenomic macronucleus contains a number of diploid subnuclei and that they differentiate into two or, rarely, more types to give a heterogeneous macronucleus. In each subunit, all but one allele is repressed. Random segregation of the subnuclei at binary fission would eventually lead to macronuclei containing only one kind of subnucleus. At that time, the specific mating type would be expressed (Allen, 1965, 1967c; Nanney and Allen, 1959; Bleyman *et al.*, 1966). A subunit number of 45 can be derived by a mathematical treatment (Schensted, 1958), and about 80 suspicious chromatin aggregates can be seen on examination of the nuclei of dividing cells (Nilsson, 1970c). Daughter nuclei receive about 40 of these structures. Direct determination of the deoxyribonucleic acid content of micro- and macronuclei gives values that are compatible with the subunit hypothesis (Flavell and Jones, 1970). However, there is now some evidence indicating that the subnuclei may be haploid rather than diploid (Woodard *et al.*, 1968; Nilsson, 1970c).

The process by which phenotypic changes occur within a heterozygous clone is referred to as allelic repression or phenotypic drift. Once the mating type is established by this process, it is stable as long as the original macronucleus exists. For an unexplained reason, starvation increases the rate of stabilization (Nanney and Caughey, 1955).

Immobilization Antigens

If cells from a given strain of *Tetrahymena pyriformis* are injected into a rabbit (Loefer *et al.*, 1958) or even a cockroach (Seaman and Robert, 1968), an antiserum is produced which acts against other cells of the same kind. When a ciliate carrying a particular antigen (serotype) is placed in fluid containing the homologous antibody, its movement ceases and the cell eventually dies. A large number of serological types are known for *T. pyriformis* (Loefer *et al.*, 1958; Elliott and Byrd, 1959), but the four serotypes of syngen 1 have been most extensively investigated. These four are dependent on temperature for expression. Serotype L occurs at 10 to 20°C, serotype H at 20 to 35°C, and serotype T at 36 to 40°C. Serotype I is expressed at room temperature by cells exposed to dilute anti-H serum and maintained in its presence (Margolin *et al.*, 1959; Nanney, 1963c; Phillips, 1967a; Juergensmeyer, 1969). Cells with the L serotype can be converted to the H serotype, and vice versa, simply by changing the temperature. Some families of cells, however, have a preference for either the H or the L serotype and maintain these for long periods of time at unfavorable temperatures (Juergensmeyer, 1969). Strain differences are known for the H and T serotypes, but the L antigen is the same regardless of strain (Nanney and Dubert, 1960; Phillips, 1967a; Juergensmeyer, 1969).

Crosses between the H and L serotypes performed at intermediate temperatures have been informative. When conjugation is abortive after cytoplasmic exchange, the cells separate and produce clones with distinctive serotypes; but when conjugation goes to completion, no systematic differences between exconjugants can be demonstrated. Such results indicate that the cytoplasm does not play a significant role in the maintenance of serotypes (Nanney, 1963c; Juergensmeyer, 1969). Instead, macronuclear control of serotype determination is indicated.

Six distinct classes of the H serotype have been found for syngen 1 (Loefer and Owen, 1961). These antigens are controlled by a single genetic locus with several alleles (Nanney and Dubert, 1960). Individual, terminal members of a particular inbred series are of a single class; early members of this series may be of more than one class (Loefer and Owen, 1961; Nanney and Dubert, 1960; Nanney *et al.*, 1964). For one H heterozygote, the critical differentiation as to what serotype will be eventually expressed occurs prior to the first division of the macronucleus; but for two other heterozygotes, the differentiation occurs later (Nanney *et al.*, 1964; Bleyman *et al.*, 1966). The selection of a single serotype during the vegetative growth following division is a process similar to, but independent of, that for mating-type selection. Some of the alleles in each macronuclear subunit are repressed during this growth, and the subunits eventually segregate out to give pure types (Nanney and Dubert, 1960; Nanney, 1963b; Nanney *et al.*, 1963).

Some anomalous serotypes have appeared in the F_2 progeny of a single cross involving alleles at the H locus (Nanney, 1962). The aberrant phenotypes disappear at conjugation and cannot be recovered in the progeny. The basis for the unusual types has not been explained.

Among eleven inbred families, there are three different T (high-temperature) phenotypes, which are controlled by three alleles at a single locus (Phillips, 1967a,b). The T heterozygotes differentiate to express only one of the parental serotypes. The kinetics of this stabilization is similar to that for the H serotypes, but the differentiation of T serotypes is independent of H serotype and mating-type differentiation (Bleyman *et al.*, 1966). Although the crosses are made at 25°C and vegetative pedigree lines are maintained at this temperature, T-serotype differentiations are accomplished and maintained in the absence of their expression (Phillips, 1967a,b). The T serotype may be the same as the high-temperature serotype S observed by Allen (1962).

The I serotype, which is expressed only in the presence of anti-H serum (Margolin *et al.*, 1959), is not a single antigenic specificity as are H and T but is instead an array of unstable antigenic types which frequently transform among themselves. Four I types are known (Juergensmeyer, 1969).

Esterases

A number of nonspecific esterases are associated with cytoplasmic granules and food vacuoles of *Tetrahymena* (Allen *et al.*, 1963a) and their activity increases as the cultures age (Fennell and Pastor, 1958; Koehler and Fennell, 1964). The control of two different esterase groups has been analyzed genetically. The E-1 group consists of twelve or more isozymes which are distinguished on electrophoresis, and these isozymes can be associated with one of two subgroups (Allen, 1960, 1961a, 1964b). One subgroup of six or more isozymes is found in one homozygote, and the remaining subgroup of six or more is in a second homozygote. Both subgroups are present in the heterozygote, and no new isozymes appear. This evidence indicates that the E-1 isozymes are controlled by codominant alleles at a single locus. The E-1 locus is loosely linked with the mating-type (*mt*) locus—the only case of linkage known for *Tetrahymena* (Allen, 1961b, 1964a).

As for other genetic loci in syngen 1, differentiation occurs in heterozygous clones maintained for many generations. Some cells develop phenotypes like the parental homozygotes during the process of allelic repression and subsequent segregation of macronuclear subunits (Allen, 1967c, 1968). Genetic studies of the esterases are complicated by the fact that different isozymes within a homozygote appear at different times in the life history of a culture and that the isozymes are not all found in the same cellular fraction (Allen, 1961c).

Some biochemical properties of the E-1 esterase isozymes are known. The isozymes are relatively stable; resist denaturation with high concentrations of urea; preferentially cleave α-naphthylacetate and propionate; are activated by cholate, taurocholate, and Triton X–100; and are inhibited by eserine and mercaptoethanol (Fennell and Pastor, 1958; Cecil, 1959; Allen, 1960, 1965, 1968; Allen *et al.*, 1965).

The E-2 esterases are less complex; only two isozymes are present (Allen, 1961a). One isozyme is in one homozygote, the other in a second homozygote, and both are present in the heterozygote. Thus, E-2 esterases are also controlled by codominant alleles at a single locus; and, similarly to E-1, allelic repression occurs in heterozygous clones maintained for many divisions. Biochemical studies show that the E-2 isozymes are less stable than the E-1; are inactivated by urea; preferentially split α-naphthylbutyrate and valerate; are not affected by eserine and cholate; and are inhibited by taurocholate and Triton X–100 (Allen, 1960, 1965, 1968; Allen *et al.*, 1965).

Acid Phosphatases

There are many acid phosphatases in *Tetrahymena pyriformis* (Allen *et al.*, 1963a), but only those under the control of the phosphatase-1 (P-1) locus have been examined genetically. Studies are complicated by the fact that the activity of these isozymes is very low under conditions of cell growth that prevent lysosomes from forming (Allen, 1968). But by using proper growth conditions, it has been found that the P-1 isozymes are controlled by codominant alleles at a single locus and that each homozygote has one principal isozyme (either isozyme 1 or 5). In the heterozygote, as many as five isozymes (1, 2, 3, 4, and 5) may be present (Allen *et al.*, 1963b; Allen, 1968). Although all five isozymes have been found in one cell line, usually fewer are present.

There are three stable P-1 cell types—the two homozygotes and one other (with isozyme 3). Four unstable cell types have isozymes 1, 3, and 5; 1, 2, and 3; 3, 4, and 5; and 1, 2, 3, 4, and 5 (Allen, 1965, 1967c). Through the process of allelic repression, all three stable types can be derived from the clone containing all the isozymes, but only two of the three stable cell types can be derived from a clone with three isozymes. Each of the clones with three isozymes gives rise to a different combination of stable types. For instance, the clone containing isozymes 3, 4, and 5 gives rise to stable types containing either isozyme 3 or 5. This information has been interpreted to mean that the isozymes exist as dimers of three basic types (AA, BB, and AB), which can coalesce in all possible ways (five) to form tetramers. Each of seven cell types is viewed as containing either one, three, or five of the types of tetramers (Allen, 1967c, 1968).

Polymorphism

All cells of a species of *Tetrahymena*, even those in the same culture, are not alike. The cells have irregular sizes, shapes, and arrangements of cilia; but these variations usually are reversible and are not genetically determined. Instead, epigenetic factors are involved. An outstanding irregularity is the presence of large- and small-mouth forms of the same ciliate. Those with small mouths (microstomes) feed on bacteria, but those with large mouths (macrostomes) ingest larger prey and even cannibalize. When microstomes are exposed to large prey, they develop new, enlarged mouth parts (Kidder *et al.*, 1940; Corliss, 1953), the process occurring even when the food source is separated by a membrane (Claff, 1947; Buhse, 1962) A ribonuclease-sensitive principle called "stomatin" is released by *T. pyriformis*; and it causes synchronous formation of macrostomes in *T. vorax*, a predator of *T. pyriformis* (Buhse, 1967; Buhse and Cameron, 1968). The formation of macrostomes from microstomes requires ribonucleic acid synthesis, for the process is inhibited by actinomycin D and by 2-mercapto-1-(β-4-pyridethyl)benzimidazole (MPB), which are two selective inhibitors of ribonucleic acid formation (Buhse and Nicolette, 1970; Nicolette *et al.*, 1971). The inhibition by MPB but not actinomycin D is reversed by washing the cells. A microstome becomes a macrostome by developing an entirely new mouth (Stone, 1963); but a macrostome can be partially resorbed to form a microstome (Buhse, 1962). The structural details of macrostome formation have been investigated for *T. vorax* (Buhse, 1966a,b, 1968; Buhse *et al.*, 1970) and *T. patula* (Miller and Stone, 1963).

Tetrahymena vorax can produce cysts; and when it does, all three forms, microstome, macrostome, and cyst, are interconvertible (Williams, 1960, 1961). The interconversion of microstome and macrostome can occur even in axenic cultures of *T. vorax* and *T. patula* (Williams, 1958, 1960, 1961; Buhse, 1963; Stone, 1963). At the start of such a culture, macrostomes revert to microstomes; but the cells revert again in the stationary phase (Williams, 1958). At lower temperatures, the reversion is more complete (Stone, 1963). There is an antigenic difference between microstomes and macrostomes of the same type, which suggests that different genes are expressed in the two forms (Shaw, 1960; Shaw and Williams, 1963).

REFERENCES

Abdel-Hameed, F. (1969). *J. Protozool.* **16,** Suppl. 9.
Alfert, M., and Goldstein, N. (1955). *J. Exp. Zool.* **130,** 403.
Allen, S. L. (1958). *J. Protozool.* **5,** Suppl. 12.
Allen, S. L. (1960). *Genetics* **45,** 1051.
Allen, S. L. (1961a). *Ann. N.Y. Acad. Sci.* **94,** 753.
Allen, S. L. (1961b). *Genetics* **46,** 847.
Allen, S. L. (1961c). *Amer. Zool.* **1,** 338.
Allen, S. L. (1962). *Amer. Zool.* **2,** 502.
Allen, S. L. (1963). *J. Protozool.* **10,** 413.
Allen, S. L. (1964a). *Genetics* **49,** 617.
Allen, S. L. (1964b). *J. Exp. Zool.* **155,** 349.
Allen, S. L. (1965). *Brookhaven Symp. Biol.* **18,** 27.
Allen, S. L. (1967a). *Science* **155,** 575.
Allen, S. L. (1967b). *Genetics* **55,** 797.
Allen, S. L. (1967c). *In* "Chemical Zoology" (M. Florkin and B. T. Scheer, eds.), Vol. 1, p. 617. Academic Press, New York.
Allen, S. L. (1968). *Ann. N.Y. Acad. Sci.* **151,** 190.
Allen, S. L., Misch, M. S., and Morrison, B. M. (1963a). *J. Histochem. Cytochem.* **11,** 706.
Allen, S. L., Misch, M. S., and Morrison, B. M. (1963b). *Genetics* **48,** 1635.
Allen, S. L., Allen, J. M., and Licht, B. M. (1965). *J. Histochem. Cytochem.* **13,** 434.
Allen, S. L., File, S. K., and Koch, S. W. (1967). *Genetics* **55,** 823.
Bleyman, L. K., Simon, E. M., and Brosi, R. (1966). *Genetics* **54,** 277.
Buhse, H. E., Jr. (1962). *J. Protozool.* **9,** Suppl. 15.
Buhse, H. E., Jr. (1963). *J. Protozool.* **10,** Suppl. 16.
Buhse, H. E., Jr. (1966a). *Trans. Amer. Microsc. Soc.* **85,** 305.
Buhse, H. E., Jr. (1966b). *J. Protozool.* **13,** 429.
Buhse, H. E., Jr. (1967). *J. Protozool.* **14,** 608.

Buhse, H. E., Jr. (1968). *J. Protozool.* **15,** Suppl. 10.
Buhse, H. E., Jr., and Cameron, I. L. (1968). *J. Exp. Zool.* **169,** 229.
Buhse, H. E., Jr., and Nicolette, J. A. (1970). *J. Protozool.* **17,** Suppl. 12.
Buhse, H. E., Jr., Corliss, J. O., and Holsen, R. C. (1970). *Trans. Amer. Microsc. Soc.* **89,** 328.
Cecil, J. T. (1959). *J. Protozool.* **6,** Suppl. 23.
Charret, R. (1969). *Exp. Cell Res.* **54,** 353.
Claff, C. L. (1947). *Biol. Bull.* **93,** 216.
Clark, G. M., and Elliott, A. M. (1956). *J. Protozool.* **3,** 181.
Corliss, J. O. (1952). *C. R. Acad. Sci. Ser.* **D235,** 399.
Corliss, J. O. (1953). *Parisitology* **43,** 49.
Ducoff, H. S. (1956). *J. Protozool.* **3,** Suppl. 3.
Elliott, A. M. (1959). *Annu. Rev. Microbiol.* **13,** 79.
Elliott, A. M. (1970). *J. Protozool.* **17,** 162.
Elliott, A. M., and Byrd, J. R. (1959). *J. Protozool.* **6,** Suppl. 19.
Elliott, A. M., and Clark, G. M. (1956a). *J. Protozool.* **3,** Suppl. 32.
Elliott, A. M., and Clark, G. M. (1956b). *Cytologia* **22,** 355.
Elliott, A. M., and Clark, G. M., (1957). *Biol. Bull.* **113,** 344.
Elliott, A. M., and Clark, G. M. (1958a). *J. Protozool.* **5,** 235.
Elliott, A. M., and Clark, G. M. (1958b). *J. Protozool.* **5,** 240.
Elliott, A. M., and Gruchy, D. F. (1952). *Biol. Bull.* **103,** 301.
Elliott, A. M., and Hayes, R. E. (1953). *Biol. Bull.* **105,** 269.
Elliott, A. M., and Hayes, R. E. (1954). *J. Protozool.* **1,** Suppl. 2.
Elliott, A. M., and Hayes, R. E. (1955a). *J. Protozool.* **2,** Suppl. 8.
Elliott, A. M., and Hayes, R. E. (1955b). *J. Protozool.* **2,** 75.
Elliott, A. M., and Kennedy, J. R. (1962). *Trans. Amer. Microsc. Soc.* **81,** 300.
Elliott, A. M., and Nanney, D. L. (1952). *Science* **116,** 33.
Elliott, A. M., and Tremor, J. W. (1958). *J. Biophys. Biochem. Cytol.* **4,** 839.
Elliott, A. M., Addison, M. A., and Carey, S. E. (1962). *J. Protozool.* **9,** 135.
Elliott, A. M., Studier, M. A., and Work, J. A. (1964). *J. Protozool.* **11,** 370.
Fennell, R. A., and Pastor, E. P. (1958). *J. Morphol.* **103,** 187.
Flavell, R. A., and Jones, I. G. (1970). *Biochem. J.* **116,** 155.
Gorovsky, M. A. (1965). *J. Cell Biol.* **27,** 37A.
Gruchy, D. F. (1955). *J. Protozool.* **2,** 178.
Hurst, D. D. (1957). *J. Protozool.* **4,** Suppl. 17.
Juergensmeyer, E. B. (1969). *J. Protozool.* **16,** 344.
Kidder, G. W., Lilly, D. M., and Claff, C. L. (1940). *Biol. Bull.* **78,** 9.
Koehler, L. D., and Fennell, R. A. (1964). *J. Morphol.* **114,** 209.
Lee, Y. C., and Byfield, J. E. (1969). *J. Cell Biol.* **43,** 78A.
Loefer, J. B., and Owen, R. D. (1961). *J. Protozool.* **8,** 387.
Loefer, J. B., Owen, R. D., and Christenssen, E. (1958). *J. Protozool.* **5,** 209.
McDonald, B. B. (1959). *J. Protozool.* **6,** Suppl. 17.
McDonald, B. B. (1966). *J. Protozool.* **13,** 277.
Margolin, P., Loefer, J. B., and Owen, R. D. (1959). *J. Protozool.* **6,** 207.
Miller, O. L., Jr., and Stone, G. E. (1963). *J. Protozool.* **10,** 280.
Nanney, D. L. (1953). *Biol. Bull.* **105,** 133.
Nanney, D. L. (1956). *Amer. Natur.* **90,** 291.
Nanney, D. L. (1957). *Genetics* **42,** 137.
Nanney, D. L. (1958). *Cold Spring Harbor Symp. Quant. Biol.* **23,** 327.

Nanney, D. L. (1959a). *J. Protozool.* **6,** 171.
Nanney, D. L. (1959b). *Genetics* **44,** 1173.
Nanney, D. L. (1960). *Physiol. Zool.* **33,** 146.
Nanney, D. L. (1962). *J. Protozool.* **9,** 485.
Nanney, D. L. (1963a). *Genetics* **48,** 737.
Nanney, D. L. (1963b). *In* "Biological Organization at the Cellular and Supercellular Level" (R. C. J. Harris, ed.), p. 91. Academic Press, New York.
Nanney, D. L. (1963c). *J. Protozool.* **10,** 152.
Nanney, D. L. (1968). *Annu. Rev. Genet.* **2,** 121.
Nanney, D. L., and Allen, S. L. (1959). *Physiol. Zool.* **32,** 221.
Nanney, D. L., and Caughey, P. A. (1953). *Proc. Nat. Acad. Sci. U.S.* **39,** 1057.
Nanney, D. L., and Caughey, P. A. (1955). *Genetics* **40,** 388.
Nanney, D. L., and Dubert, J. M. (1960). *Genetics* **45,** 1335.
Nanney, D. L., and Nagel, M. J. (1964). *J. Protozool.* **11,** 465.
Nanney, D. L., Caughey, P. A., and Tefankjian, A. (1955). *Genetics* **40,** 668.
Nanney, D. L., Reeve, S. J., Nagel, J., and DePinto, S. (1963). *Genetics* **48,** 803.
Nanney, D. L., Nagel, M. J., and Touchberry, R. W. (1964). *J. Exp. Zool.* **155,** 25.
Nicolette, J. A., Buhse, H. E., and Robin, M. S. (1971). *J. Protozool.* **18,** 87.
Nilsson, J. R. (1970a). *J. Protozool.* **17,** Suppl. 9.
Nilsson, J. R. (1970b). *C. R. Trav. Lab. Carlsberg* **37,** 285.
Nilsson, J. R. (1970c). *J. Protozool.* **17,** 539.
Orias, E. (1957). *J. Protozool.* **4,** Suppl. 20.
Orias, E. (1958). *J. Protozool.* **5,** Suppl. 17.
Orias, E. (1960). *J. Protozool.* **7,** 64.
Orias, E. (1963). *Genetics* **48,** 1509.
Outka, D. E. (1961). *J. Protozool.* **8,** 179.
Phillips, R. B. (1967a). *Genetics* **56,** 667.
Phillips, R. B. (1967b). *Genetics* **56,** 683.
Phillips, R. B. (1969). *Genetics* **63,** 349.
Phillips, R. B. (1971). *J. Protozool.* **18,** 163.
Ray, C. (1956a). *J. Protozool.* **3,** Suppl. 3.
Ray, C. (1956b). *J. Protozool.* **3,** 88.
Ray, C., and Elliott, A. M. (1954). *Anat. Rec.* **120,** 228.
Roth, L. E., and Minick, O. T. (1958). *J. Protozool.* **5,** Suppl. 22.
Schensted, I. V. (1958). *Amer. Natur.* **92,** 161.
Schooley, C. N. (1958). *J. Protozool.* **5,** Suppl. 24.
Seaman, G. R., and Robert, N. L. (1968). *Science* **161,** 1359.
Shaw, R. F. (1960). *J. Protozool.* **7,** Suppl. 14.
Shaw, R. F., and Williams, N. E. (1963). *J. Protozool.* **10,** 486.
Sonneborn, T. M. (1947). *Advan. Genet.* **1,** 263.
Stone, G. E. (1963). *J. Protozool.* **10,** 74.
Wells, C. (1961). *J. Protozool.* **8,** 284.
Williams, N. E. (1958). *J. Protozool.* **5,** Suppl. 11.
Williams, N. E. (1960). *J. Protozool.* **7,** 10.
Williams, N. E. (1961). *J. Protozool.* **8,** 403.
Wolfe, J. (1967). *Chromosoma* **23,** 59.
Woodard, J., Gorovsky, M., and Kaneshiro, E. (1968). *J. Cell Biol.* **39,** 182A.

CHAPTER 8

Vitamin and Inorganic Requirements

Introduction

Generally, vitamins can be considered as dietary organic compounds, the absence of which leads to deficiency diseases in animals. As related to *Tetrahymena*, however, this term is difficult to define. A proper definition should exclude the carbon source (e.g., glucose), the amino acids, the purines, the pyrimidines, and the inorganic ions of the growth medium but should include a number of compounds described as vitamins for man and also lipoate, a growth factor required only by the ciliates *Tetrahymena*, *Glaucoma*, and *Colpidium* (Holz, 1964). The vitamin requirements for *Tetrahymena*, as determined by thorough study, are not as extensive as for mammals, for neither vitamin B_{12}, vitamin C, inositol, choline, nor any of the fat-soluble vitamins (vitamins A, D, E, and K) are necessary. The vitamins needed for growth of two other ciliates, *Paramecium* and *Glaucoma*, are essentially the same as those for *Tetrahymena* (Holz, 1964). The known inorganic ion requirements for *Tetrahymena* are not as expansive as for mammals.

Thiamine and Lipoate (Thioctic Acid)

More than 30 years ago, thiamine (vitamin B_1) was established as a growth factor for *Tetrahymena (Glaucoma) pyriformis* (Lwoff and Lwoff, 1937). The organism would not grow in unsupplemented silk peptone media, but would grow if thiamine was added. Neither the pyrimidine nor the thiazole portion of the molecule would suffice. However, some other early investigations tended to show that this species could grow in the absence of thiamine, provided that certain factors from crude extracts were present (Kidder and Dewey, 1942, 1944, 1945). Later, when chemically defined media were available, it became apparent that the organism did, indeed, require thiamine (Kidder and Dewey, 1951). The requirement also applies to *T. corlissi* (Holz *et al.*, 1961a), *T. paravorax* (Holz *et al.*, 1961b), and *T. setifera* (Holz *et al.*, 1962). It is reported that certain mutant strains of *T. pyriformis* do not require this vitamin (Elliott *et al.*, 1962), but it appears necessary to await confirmation of such freedom. Addition of neopyrithiamine, an analog of thiamine, to the medium results in competitive growth inhibition (Kidder and Dewey, 1951).

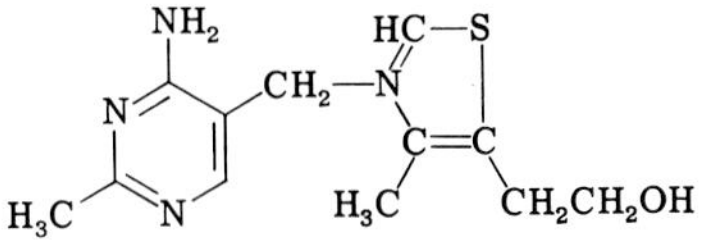

Thiamine

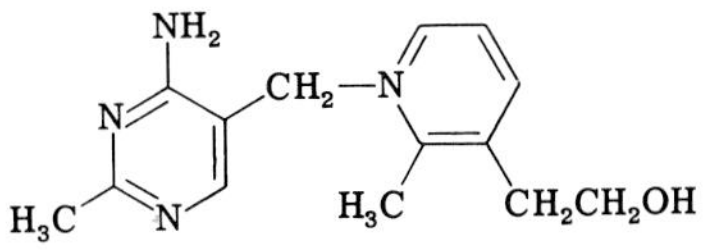

Neopyrithiamine

In the process of determining the nutritional requirements of *T. pyriformis*, Dewey (1944) discovered that extracts of natural substances contained a compound (Factor II), distinct from other vitamins, which was necessary for growth of *T. pyriformis*. Stokstad *et al.* (1949) succeeded in concentrating the compound, at that time designated Factor IIA, and proposed that the name "protogen" be used until the chemical structure was determined. Shortly thereafter, Snell and Broquist (1949) presented evidence indicating that protogen was identical to the acetate-replacing factor for *Lactobacillus casei* and to the pyruvate oxidation factor from yeast. Reed *et al.* (1951) succeeded in crystallizing the growth factor and proposed that it be named α-lipoic acid. The name was suggested because of its solubility in organic solvents and the prefix assigned because it was thought that several varieties of the vitamin would be found. Another group (Patterson *et al.*, 1951) crystallized an active form of the factor and named it "protogen B." This form proved to be the monosulfoxide (Reed *et al.*, 1951). Other forms were mixed disulfides of α-lipoic acid with naturally occurring thiols (Reed *et al.*, 1951). The precise chemical structure of α-lipoic acid was found to be, in the reduced form, 6,8-dithiooctanoic acid (Bullock *et al.*, 1952):

$$\underset{\text{SH}}{\underset{|}{CH_2}}-CH_2-\underset{\text{SH}}{\underset{|}{CH}}-CH_2CH_2CH_2CH_2COOH$$

Lipoic (thioctic) acid

In the oxidized form (1,2-dithiolane-3-valeric acid) a disulfide bond is present.

It was noted early (Kidder *et al.*, 1950) that acetate reduced the lipoate requirement of *T. pyriformis* and that cells grown in media without acetate and containing suboptimal amounts of lipoate accumulated pyruvate, α-ketoglutarate, and α-ketoisocaproate in sufficient quantity to be lethal (Dewey and Kidder, 1953a). Cells grown with inadequate amounts of thiamine also accumulate α-keto acids (Tittler *et al.*, 1952; Dewey and Kidder, 1953a). Seaman (1952) found that lipoate was required for the formation of acetyl CoA from pyruvate and proceeded to purify partially the pyruvate dehydrogenase system (Seaman, 1953). He also observed that thiamine is required for decarboxylation of α-ketoglutarate, which accounts for the observations that α-keto acids accumulate in thiamine-deficient cultures and that the thiamine requirement is spared by fat and acetate (Tittler *et al.*, 1952). Less thiamine is needed by the cell if there is no need to convert pyruvate to acetate. Lipoate can be removed from enzymes by treatment with alumina (Seaman and Naschke, 1955), and such treatment decreases pyruvate oxidation. The activity can be restored by addition of the vitamin.

Lipoic acid is now recognized as a coenzyme, prosthetic group, or substrate in plants and animals. Its main function, that of oxidative decarboxylation of α-keto acids, is illustrated in Fig. 8.1.

All species and strains of *Tetrahymena* that have been studied require lipoate (Kessler, 1961; Holz *et al.*, 1961a,b; 1962; Elliott *et al.*, 1962).

Riboflavin

Hall (1944) was first to note that *Tetrahymena* require riboflavin for optimum growth. He reported that some growth could be obtained in the

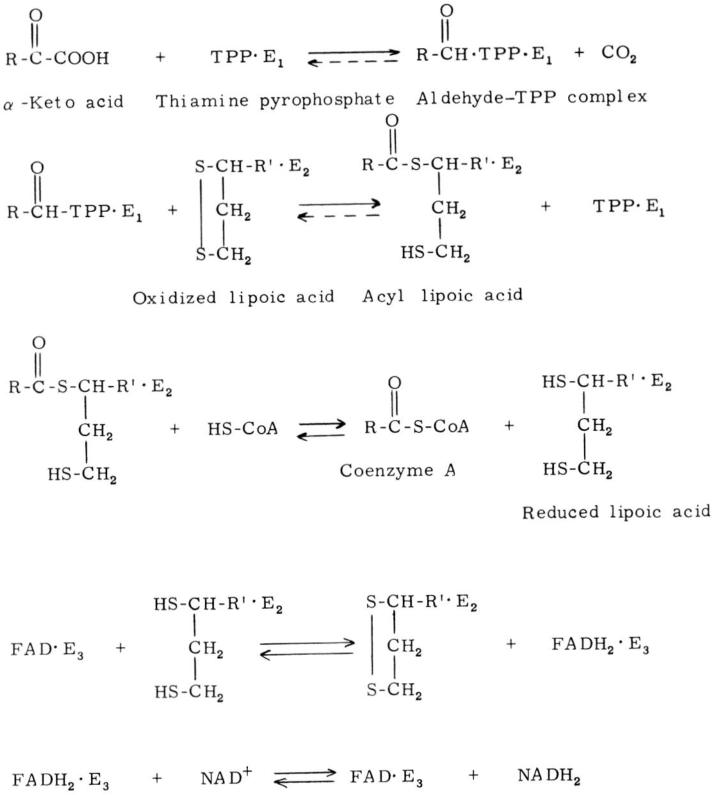

$$\mathrm{R{-}\overset{O}{\overset{\|}{C}}{-}COOH} + \mathrm{TPP{\cdot}E_1} \rightleftharpoons \mathrm{R{-}\overset{O}{\overset{\|}{C}}H{\cdot}TPP{\cdot}E_1} + \mathrm{CO_2}$$

α-Keto acid　　Thiamine pyrophosphate　　Aldehyde–TPP complex

$$\mathrm{R{-}\overset{O}{\overset{\|}{C}}H{-}TPP{\cdot}E_1} + \begin{matrix}\mathrm{S{-}CH{-}R'{\cdot}E_2}\\ \mathrm{CH_2}\\ \mathrm{S{-}CH_2}\end{matrix} \rightleftharpoons \begin{matrix}\mathrm{R{-}\overset{O}{\overset{\|}{C}}{-}S{-}CH{-}R'{\cdot}E_2}\\ \mathrm{CH_2}\\ \mathrm{HS{-}CH_2}\end{matrix} + \mathrm{TPP{\cdot}E_1}$$

Oxidized lipoic acid　　Acyl lipoic acid

$$\begin{matrix}\mathrm{R{-}\overset{O}{\overset{\|}{C}}{-}S{-}CH{-}R'{\cdot}E_2}\\ \mathrm{CH_2}\\ \mathrm{HS{-}CH_2}\end{matrix} + \mathrm{HS{-}CoA} \rightleftharpoons \mathrm{R{-}\overset{O}{\overset{\|}{C}}{-}S{-}CoA} + \begin{matrix}\mathrm{HS{-}CH{-}R'{\cdot}E_2}\\ \mathrm{CH_2}\\ \mathrm{HS{-}CH_2}\end{matrix}$$

Coenzyme A　　Reduced lipoic acid

$$\mathrm{FAD{\cdot}E_3} + \begin{matrix}\mathrm{HS{-}CH{-}R'{\cdot}E_2}\\ \mathrm{CH_2}\\ \mathrm{HS{-}CH_2}\end{matrix} \rightleftharpoons \begin{matrix}\mathrm{S{-}CH{-}R'{\cdot}E_2}\\ \mathrm{CH_2}\\ \mathrm{S{-}CH_2}\end{matrix} + \mathrm{FADH_2{\cdot}E_3}$$

$$\mathrm{FADH_2{\cdot}E_3} + \mathrm{NAD^+} \rightleftharpoons \mathrm{FAD{\cdot}E_3} + \mathrm{NADH_2}$$

FIG. 8.1. Function of lipoic acid in the oxidative decarboxylation of α-keto acids. E_1, pyruvate dehydrogenase; E_2, dihydrolipoyl transacetylase; E_3, dihydrolipoyl dehydrogenase; TPP, thiamine pyrophosphate; FAD, flavin adenine dinucleotide. (From Reed, 1966.)

absence of this vitamin. Kidder and Dewey (1945) in their early work had similar results, but later these workers were able to show that the requirement was absolute—a fact which contradicts reports by Elliott (1949) and by Seaman (1953). The requirement for riboflavin has been

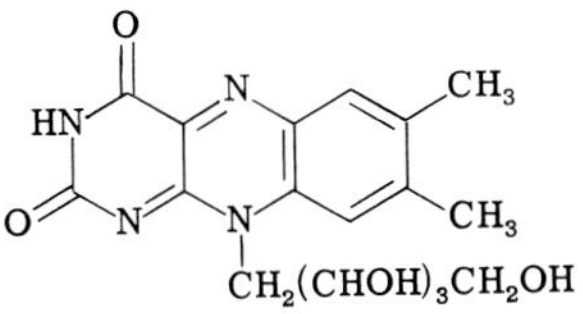

O, HN, N, CH_3, O, N, N, CH_3, $CH_2(CHOH)_3CH_2OH$

Riboflavin

extended to several species of *Tetrahymena* (Kidder and Dewey, 1951; Holz *et al.*, 1961b, 1962; Kessler, 1961; Elliott *et al.*, 1962).

Pyridoxal and Pyridoxamine

A source of pyridoxal or pyridoxamine is necessary for growth of *Tetrahymena pyriformis*. Kidder and Dewey (1949b) showed that their earlier results to the contrary were in error. The activity of pyridoxal is equal to that of pyridoxamine and is 500 times as strong as pyridoxine (Kidder and Dewey, 1951). Pyridoxal or pyridoxamine is also required by the *Tetrahymena* species *T. patula* (Kessler, 1961; Holz, 1964) and *T. vorax*

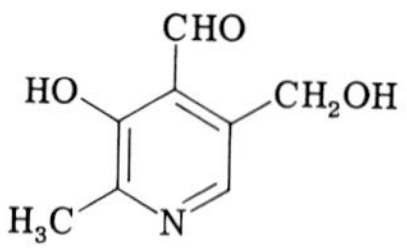

CHO, HO, CH_2OH, H_3C, N

Pyridoxal

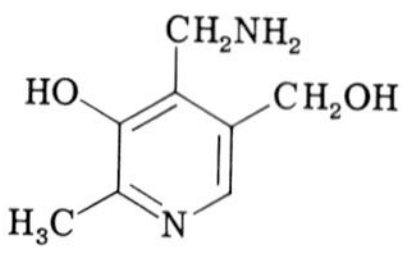

CH_2NH_2, HO, CH_2OH, H_3C, N

Pyridoxamine

(Kessler, 1961). However, *T. corlissi* (Holz, 1964; Kessler, 1961), *T. paravorax* (Holz *et al.*, 1961b), *T. setifera* (Holz *et al.*, 1962), and strains of *T. pyriformis* described as "pyridoxine mutants" are said not to require the vitamin (Elliott and Clark, 1958a,b; 1963). A reinvestigation of the nonrequiring species and strains is in order.

Pantothenate

As for a number of other vitamins, pantothenate was first viewed only as a stimulatory factor, rather than as an absolute requirement. But with the development of defined media, it became obvious that the organism could not grow in its absence (Kidder and Dewey, 1949a). All the species and strains tested show this requirement (Holz *et al.*, 1961a, 1962; Kessler, 1961; Seaman, 1952; Elliott *et al.*, 1962). *Tetrahymena pyriformis* cannot readily utilize coenzyme A (Kidder and Dewey, 1951); pantothenate is best, followed by pantetheine and pantetheine phosphate (Dewey and Kidder, 1954).

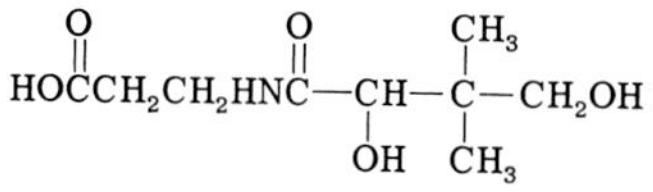

$$HOC(=O)CH_2CH_2HNC(=O)\text{—}CH(OH)\text{—}C(CH_3)_2\text{—}CH_2OH$$

Pantothenic acid

$$HSCH_2CH_2HNC(=O)CH_2CH_2HNC(=O)\text{—}CH(OH)\text{—}C(CH_3)_2\text{—}CH_2OH$$

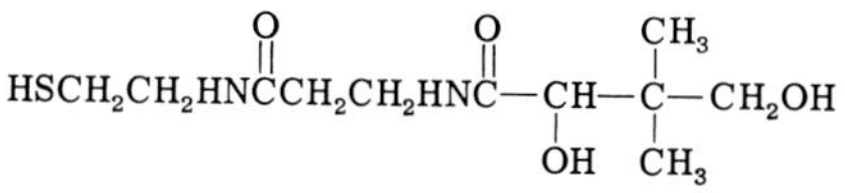

Pantetheine

Sulfonylureas may interfere with pantothenate metabolism. Several such drugs do not inhibit the growth of cells in media with the optimal concentration of pantothenate, but, at lower levels of the vitamin, they are inhibitory (Shenoy and McLaughlan, 1959).

Folate (Pteroylglutamate)

Early studies on the vitamin requirements showed that folate is necessary for the growth of *Tetrahymena* (Kidder, 1946); all species and strains studied have this requirement (Holz *et al.*, 1961a,b; 1962; Kessler, 1961; Seaman, 1952; Elliott *et al.*, 1962). Tested in the defined medium, the mono- and polyglutamates of folic acid have equal activity (Kidder and Dewey, 1951); but as evidenced by its inability to utilize fragments of the folic acid molecule, the organism does not possess the enzymes necessary for joining the pteridine portion of the molecule to the *p*-aminobenzoic acid or for joining pteroic acid to glutamate (Kidder and Dewey, 1947).

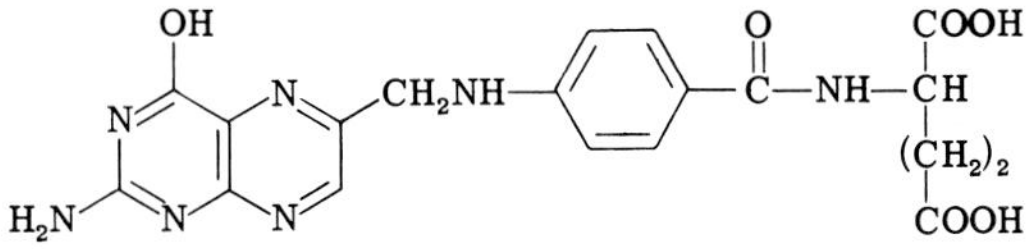

Folic acid

There has appeared a report that 4-aminopteroylglutamate (aminopterin), an inhibitory folate analog for mammals, can actually be utilized for growth by *T. pyriformis* (Kidder *et al.*, 1951a), implying that the ciliate is capable of converting it to folate by deamination. The report also states that N^{10}-methylpteroylglutamate (methopterin) can be utilized for growth. However, later results showed that the sample of aminopterin was grossly contaminated with folate, thus nullifying the previous conclusions (Dewey and Kidder, 1953b). The preparation of methopterin was also likely contaminated. 4-Amino-N^{10}-methylpteroylglutamate (amethopterin) cannot be utilized and is inhibitory to growth (Kidder *et al.*, 1951a). A combination of amethopterin and uridine inhibits the second synchronous division of heat-treated *T. pyriformis.* This combination apparently produces thymidine starvation in the organism, for the inhibition can be overcome by thymidine (Zeuthen, 1968).

The folate requirement of *T. pyriformis* is moderately spared by thymidine or deoxycytidine. There is also some sparing activity by glycine, *p*-aminobenzoic acid, creatine, and vitamin B_{12} (Dewey and Kidder, 1953b; Erwin, 1960). To a limited extent, thymidine, thymidine monophosphate

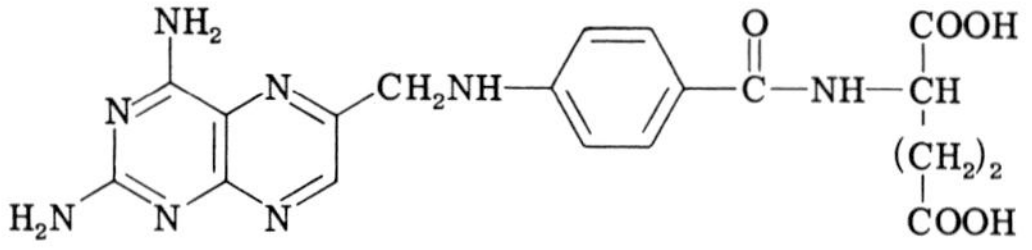

Aminopterin

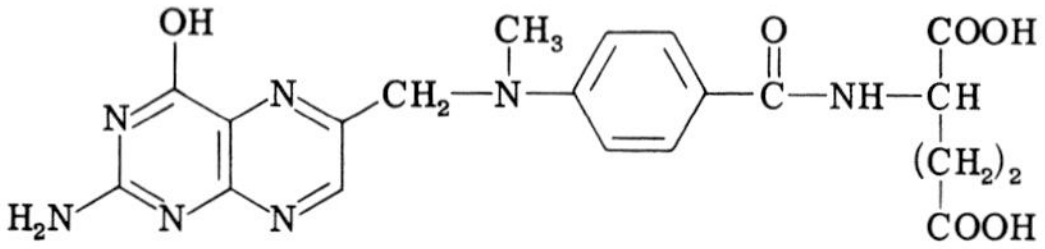

Methopterin

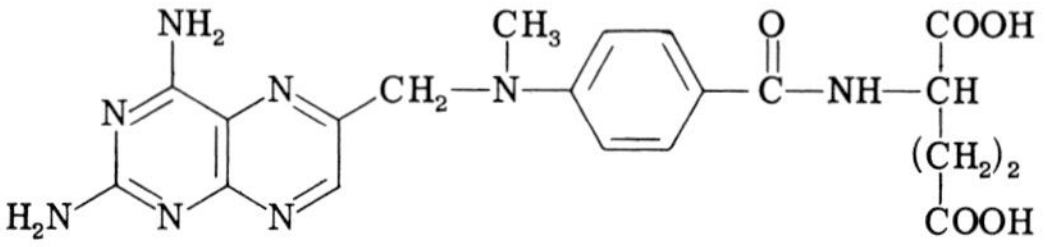

Amethopterin

(TMP), 5-methyldeoxycytidine, and 5-methyldeoxycytidine monophosphate reverse the inhibition of growth in media deficient in folate (Wykes and Prescott, 1968).

Yet another report states that some substituted pyrimidines can spare the folate requirement of *T. pyriformis* (Kidder and Dewey, 1961). Those that are effective are 2,5,6-triamino-4-hydroxypyrimidine, 2,6-diamino-4-hydroxy-5-formylaminopyrimidine, and 2,6-diamino-4-hydroxy-5-acetylaminopyrimidine. These sparing effects were at first interpreted to mean that the organism utilizes the pyrimidine derivatives for the synthesis of an

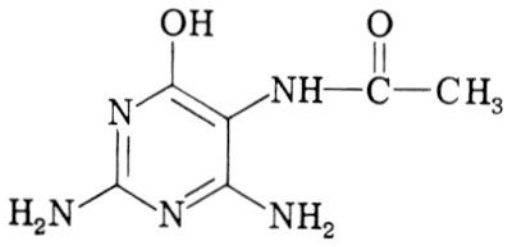

2, 6-Diamino-4-hydroxy-5-acetylaminopyrimidine

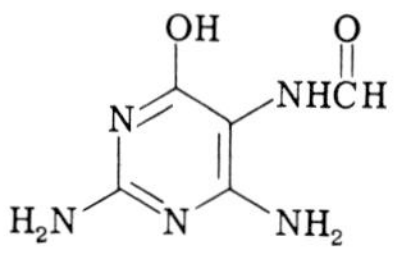

2, 6-Diamino-4-hydroxy-5-formylaminopyrimidine

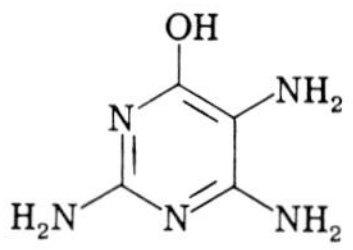

2, 5, 6-Triamino-4-hydroxypyrimidine

unconjugated pteridine, which could also be derived from folate when this is the only available source. However, later results showed that labeled folate could not give rise to unconjugated pteridines (Kidder and Dewey, 1968). The basis for the sparing effects by the pyrimidines is not known.

A search for an unconjugated pteridine in *T. pyriformis* began after it was shown that this organism produces a compound that replaces biopterin as a growth factor for *Crithidia*. Subsequently, a new pteridine, unique to *Tetrahymena*, was isolated from this organism (Kidder and Dewey, 1968). Evidence indicates that this compound is ciliapterin [D- or L-2-amino-4-hydroxy-6-(*threo*-dihydroxypropyl)pteridine] and that it is produced entirely from guanosine. It differs from biopterin in the configuration of the side chain. There is some apparent dissention regarding the structure of the substance, for the synthetic L-*threo* compound differs from ciliapterin in its biological and chromatographic properties (Guroff and Rhoads, 1969). A D-*threo* configuration for ciliapterin could account for the differing biological properties, but the chromatographic properties of the D and L isomers should be the same.

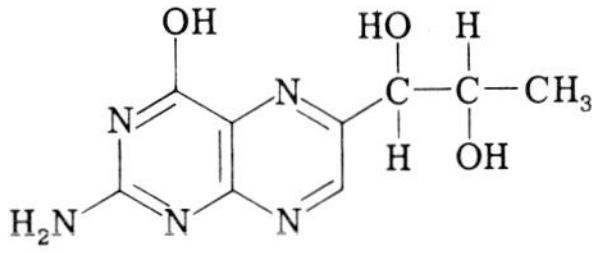

Ciliapterin

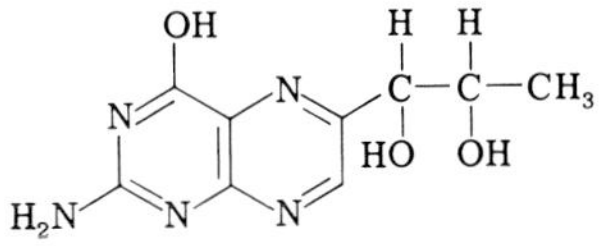

Biopterin

The function of ciliapterin remains obscure, but it is postulated that an unconjugated pteridine is involved both in the desaturation of fatty acids and in the cyclization of squalene to form sterol-like compounds (Dewey and Kidder, 1964). Whether ciliapterin is active as a cofactor in the phenylalanine and tyrosine-hydroxylating systems remains to be determined.

No hydroxymethylpteridine can be detected in extracts of the organism, indicating its inability to produce this compound, a precursor of folate. A block in folate synthesis can be visualized as starting with the lack of an enzyme cleaving the 3-carbon side chain to form the hydroxymethyl derivative (Kidder and Dewey, 1968). The remainder of the biosynthetic pathway leading to folate is probably absent (Fig. 8.2).

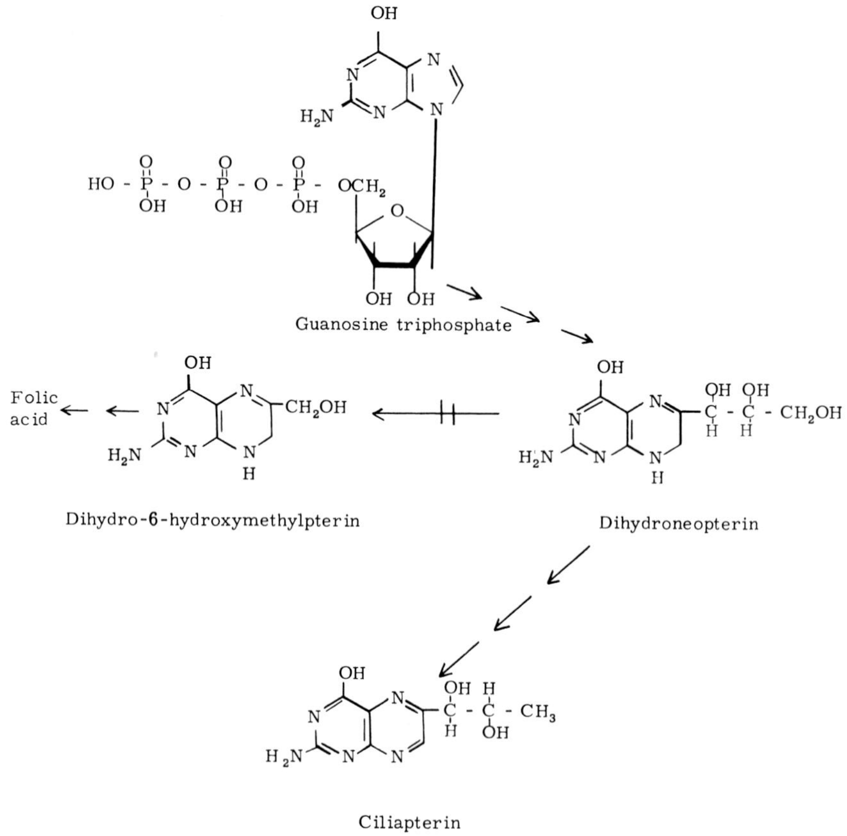

FIG. 8.2. Pteridine metabolism in *Tetrahymena*.

Niacin

When grown in the defined medium, *Tetrahymena pyriformis* has an absolute requirement for niacin (Kidder and Dewey, 1949b; Elliott, 1949), which is not spared by high levels of tryptophan (Kidder *et al.*, 1949). The requirement is also known for the *Tetrahymena* species *T. corlissi*, *T. paravorax*, and *T. setifera* (Holz *et al.*, 1961a,b, 1962).

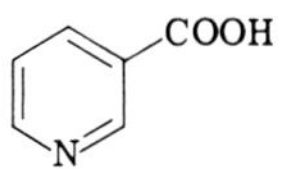

Niacin

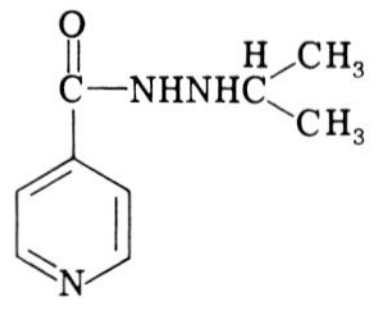

Iproniazid

Niacin and nicotinamide adenine dinucleotide (NAD) reverse inhibition of growth of *T. pyriformis* caused by the teratogenic tranquilizer, thalidomide, suggesting that the developmental abnormalities may be associated with inhibited respiration (Frank *et al.*, 1963). High concentrations of iproniazid, an analog of niacin, inhibit growth and respiration of the ciliate (Krezanowski, 1961).

Biotin

This vitamin is required by *Tetrahymena corlissi*, *Tetrahymena paravorax*, and *Tetrahymena setifera* (Holz *et al.*, 1961a,b, 1962) and is possibly necessary for the growth of *Tetrahymena pyriformis*, although no absolute requirement can be demonstrated for this species (Kidder and Dewey,

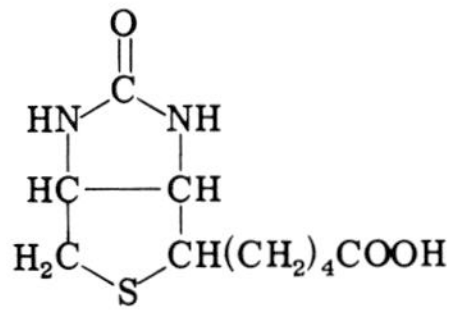

$$\begin{array}{ccc} & O & \\ & \| & \\ HN & C & NH \\ | & & | \\ HC & — & CH \\ | & & | \\ H_2C & & CH(CH_2)_4COOH \\ & S & \end{array}$$

Biotin

1949a). Neither avidin nor desthiobiotin is inhibitory. Elliott *et al.*(1964) have reported that some strains of *T. pyriformis* isolated from the Pacific Ocean require biotin.

Vitamin B_{12}

No ciliate is known to require a member of the vitamin B_{12} group. However, *Tetrahymena corlissi* and *Tetrahymena pyriformis* produce a vitamin B_{12} compound which can support the growth of the bacteria *Escherichia coli* and *Lactobacillus leichmannii* but not that of the protozoans *Ochromonas malhamensis* and *Euglena gracilis* (Erwin, 1962).

Vitamin E

A search for α-tocopherol, a form of vitamin E, in extracts of *Tetrahymena pyriformis* was unsuccessful (Green *et al.*, 1959).

Biochemical Assays

The vitamin requirements of *Tetrahymena* have allowed this organism to become a sensitive tool in the bioassay of these compounds (Hutner *et al.*, 1959; Baker *et al.*, 1960a,b; Baker and Sobotka, 1962; Aaronson *et al.*, 1964). The requirements have also been used to evaluate the effects of γ-radiation on the vitamins essential for growth (Elliott *et al.*, 1954).

Caution is required during biochemical assays, for riboflavin, thiamine, and folate are sensitive to light. The inhibitory effect of strong, visible light on the growth of *Tetrahymena* is due to the destruction of these vitamins in the medium (Dewey and Kidder, 1953b; Phelps, 1959; Lee, 1969). The organism reportedly can be trained to avoid light (Bergstrom, 1969a,b).

Water

Water, an obvious necessity, does not flow freely through the cellular membrane of *Tetrahymena*. It is not known whether this is due to a limited number of sites in the membrane or to restricted water movement through the membrane as a whole (Prescott and Zeuthen, 1953).

Sodium, Potassium, and Chloride Ions

Tetrahymena pyriformis requires potassium ions for growth (Kidder *et al.*, 1951b). It probably also requires sodium and chloride ions, but dependence on these ions cannot be demonstrated due to the extreme difficulty in preparing sodium-free and chloride-free media. The ciliate maintains a high intracellular level of potassium ions and a lower level of sodium ions (Dunham and Child, 1961). Table 8.1 shows the intracellular and extracellular concentrations of these ions in a proteose–peptone

TABLE 8.1

ION CONTENT OF *Tetrahymena pyriformis* IN 2% PROTEOSE–PEPTONE[a]

Site	Potassium (mEq/liter)	Sodium (mEq/liter)	Chloride (mEq/liter)
Cells	31.6 ± 0.4	12.7 ± 1.4	—
Medium	4.8 ± 0.1	36.5 ± 0.1	28.7 ± 0.1

[a] From Dunham and Child (1961).

medium. In this case, the potassium ion is concentrated more than sixfold. In a more dilute medium, the internal/external ratio for this ion is greater than 100, and 90% of the potassium ion content is exchangeable within 5 hr. Increasing the extracellular potassium ion content increases the internal content; but at high levels, equilibration is less rapid. Dunham and Child (1961) concluded that potassium ions are actively transported and that the inexchangeable portion is attached to internal binding sites which have a strong affinity for the ions.

For cells in a proteose–peptone medium, the internal sodium ion content is about one-third that of the internal potassium ion and one-third the sodium ion content of the medium (Table 8.1). In a dilute medium, about 60% of the sodium ion content is exchangeable (Dunham and Child, 1961), with the remainder apparently being attached to internal sites similar to those for potassium ions. Increasing the external sodium ion level up to 20 m*M* does not increase the internal level, suggesting that a sodium extrusion system is present. At higher levels, however, the extrusion system is overcome, and internal increases are noted. The chloride ion content is normally maintained at a lower level in the cells than in the environment, and this level does not increase until a saturation point is reached.

Potassium and sodium ion regulation in *T. pyriformis* has been investigated by adding agents that modify the intracellular levels (Andrus and Giese, 1963). Anoxia, iodoacetate, azide, and 2,4-dinitrophenol, all of which reduce energy metabolism, induce a 30–50% loss of potassium ions, with no reciprocal gain in the sodium ion level. This implies that the sodium and potassium ion-regulatory systems are independent. Ouabain, a glycoside which inhibits active transport of sodium in cardiac muscle, has no effect on potassium ion regulation in *T. pyriformis* (Conner, 1967). Exposure of *T. pyriformis* to low temperatures in a medium containing sodium ions produces a loss of cellular potassium and an increase in sodium ions. Normal conditions are restored when the cells are returned to 25°C. Reestablishment of the high potassium–low sodium levels does not appear to be coupled, further supporting the proposal that independent mechanisms exist for extrusion of sodium ions and entry of potassium ions (Andrus and Giese, 1963). There is a marked increase in internal sodium ion content and a decrease in potassium ion content during the final heat shock prior to cell division (Holm, 1970). Cells recovering from increased medium osmolarity take up large amounts of potassium ion if the potassium is present in the medium (Kramhøft, 1970).

Tetrahymena can adapt to a medium of much higher tonicity than that normally encountered (Chatton and Tellier, 1927; Loefer, 1939). About

2% of the cells exposed to the osmotic shock of a high sodium chloride content show a tolerance which is of a heritable nature (Dunham, 1964). A larger percentage of cells can be adapted to the hypertonic conditions by exposing them to an intermediate concentration of salt. The adapted cells are able to maintain a low sodium ion level internally.

The mechanism for maintenance of electroneutrality within the cell is unknown, but Dunham and Child (1961) show that the Gibbs–Donnan equilibrium is inadequate as an explanation.

The uptake of amino acids increases with increasing amounts of sodium ion in the medium, and the addition of L-phenylalanine to a culture causes a decrease in the intracellular sodium ion content (Hoffmann and Kramhøft, 1969). A conclusion from this information is that the sodium ion has an important role in the uptake of amino acids.

Orthophosphate

A requirement for phosphate by *Tetrahymena pyriformis* is evident (Kidder *et al.*, 1951b) and is not surprising in view of the extensive involvement of phosphate esters in metabolism. The cellular orthophosphate pool is approximately 4.9×10^{-3} *M* (Cline, 1965), but this can be reduced by the heat treatment used to induce synchrony (Hamburger and Zeuthen, 1960).

Orthophosphate is rapidly taken up by *T. pyriformis* in a manner that does not rely on food vacuole formation (Pruett, 1965). Uptake is apparently a site-mediated process, since the rate of uptake follows Michaelis–Menten kinetics and is dependent on temperature. The K_m value is 8×10^{-5} *M* at pH 7.5, and the Q_{10} (18°–28°C) is 1.7. These values are too high to represent simple diffusion of orthophosphate across the cellular membrane (Pruett, 1965; Pruett *et al.*, 1967), but the pattern does not distinguish between a membrane transport mechanism and metabolic incorporation of orthophosphate. Erickson and Conner (1968) present evidence to show that a physical phenomenon is initially involved in the accumulation of orthophosphate by the cell and that the orthophosphate is not used metabolically in the transport process. The pH of the culture fluid has a marked influence on orthophosphate accumulation, especially for low concentrations of this ion; but prior growth at various pH values has no effect. The pH optimum for accumulation is 6.5, suggesting that the monobasic form of the ion is transported more rapidly (Conner *et al.*, 1961a).

Other factors influencing the rate of orthophosphate entry into *T. pyriformis* are strain and mating type (Slater and Tremor, 1962), population density (Slater and Tremor, 1962), amount of iron in the cells (Conner and Pruett, 1963; Conner and Cline, 1964), and the presence of glucose in the medium (Conner and Pruett, 1963; Hamburger and Zeuthen, 1960).

Entry of orthophosphate into the cell may be dependent on the energy derived from glucose by glycolysis. Glucose, fructose, mannose, and D-glyceraldehyde increase the orthophosphate accumulation, but acetate decreases the rate of uptake (Slater and Tremor, 1962; Conner and Pruett, 1963; Conner, 1967). Pyruvate and ethanol, which are utilized by the cell, and a number of sugars which are not metabolized do not increase uptake (Conner, 1967). Further, the metabolic inhibitors 2,4-dinitrophenol and iodoacetate reduce the influx and increase the efflux of orthophosphate. Glucose partially overcomes the effect of 2,4-dinitrophenol but not that of the glycolytic inhibitor iodoacetate. Loss of orthophosphate induced by iodoacetate is partially reversed by acetate; but, in this case, the metabolite has no effect on entry of orthophosphate into the cell (Conner, 1967). Phenothiazine derivatives inhibit orthophosphate uptake in a manner different from that of 2,4-dinitrophenol (Rogers, 1968).

If *T. pyriformis* cells are placed in a nonnutrient medium, orthophosphate is excreted at a rate linear with time (Leboy *et al.*, 1964). The primary source of excreted phosphate is ribonucleic acid (Cline, 1965, 1966; Cline and Conner, 1966). Certain sterols influence the efflux of orthophosphate from *T. pyriformis* (Conner *et al.*, 1961b). The effect of sterols is not reversed by either 2,4-dinitrophenol or iodoacetate (Conner, 1967), leading workers in this field to change their previous opinion that the sterols affect energy metabolism (Conner *et al.*, 1961b). Nevertheless, efflux of orthophosphate seems to correlate with a low-energy state of cells.

Other Inorganic Ions

Tetrahymena pyriformis demonstrates a definite growth response in the presence of magnesium ions (Kidder *et al.*, 1951b; Slater, 1952; Hall, 1954). The magnesium requirement can also be demonstrated by the addition of oxalate or citrate, which are known to bind magnesium ions strongly.

The organism apparently does not require calcium ions (Kidder *et al.*, 1951b), for the effect of oxalate, which binds both calcium and magnesium, can be reversed by adding more magnesium. Nevertheless, calcium ions and strontium ions are accumulated by intact cells (Ballentine and Burford, 1964). Under certain conditions, the calcium and magnesium salts of inorganic pyrophosphate are precipitated in the cell to form granules (Munk and Rosenberg, 1969; Rosenberg and Munk, 1969); but in a phosphate-free medium, the granules disappear within 6 to 8 hr. Calcium ions are removed from the cells by an active, rapid process, but no magnesium leaves the cell until all the granules have disappeared. The calcium and magnesium ions may be part of the control mechanism governing the deposition and utilization of pyrophosphate (Rosenberg and Munk, 1969).

Kidder *et al.* (1951b) could not demonstrate a definite requirement for iron, but iron-deficient cells have a lowered influx of orthophosphate and a reduced oxidative capacity that can be restored by adding ferrous ions (Conner and Pruett, 1963; Conner and Cline, 1964). Furthermore, addition of inorganic iron salts to media results in marked alterations in metabolism. There is increased growth, pyridine hemochromagen concentration, conversion of heme to cytochromes, glycogen concentration, and degree of unsaturation of fatty acids (Shug *et al.*, 1969).

Tetrahymena pyriformis exhibits a definite requirement for copper ions, but not for manganese, zinc, cobalt, fluoride, borate, or molybdate (Kidder *et al.*, 1951b). Perhaps some of the latter group will be necessary in media prepared from more refined components. The organism is known to accumulate cobalt against a concentration gradient (Slater, 1957); but high levels of cobalt and nickel are toxic, possibly by inhibiting an intracellular ribonuclease (Roth, 1956).

REFERENCES

Aaronson, S., Baker, H., Bensky, B., Frank, O., and Zahalsky, A. C., (1964). *Develop. Ind. Microbiol.* **6,** 48.

Andrus, W. DeW., and Giese, A. C. (1963). *J. Cell. Comp. Physiol.* **61,** 17.

Baker, H., and Sobotka, H. (1962). *Advan. Clin. Chem.* **5,** 173.

Baker, H., Frank, O., Pasher, I., Dinnerstein, A., and Sobotka, H., (1960a). *Clin. Chem.* **6,** 36.

Baker, H., Frank, O., Pasher, I., Hutner, S. H., and Sobotka, H. (1960b). *Clin. Chem.* **6,** 572.

Ballentine, R., and Burford, D. D. (1964). *Life Sci.* **3,** 1455.

Bergstrom, S. R. (1969a). *Scand. J. Psychol.* **10,** 16.

Bergstrom, S. R. (1969b). *Scand. J. Psychol.* **10,** 81.

Bullock, M. W., Brockman, J. A., Patterson, E. L., Pierce, J. V., and Stokstad, E. L. R. (1952). *J. Amer. Chem. Soc.* **74,** 3455.

Chatton, E., and Tellier, L. (1927). *C. R. Soc. Biol.* **97,** 780.

Cline, S. G. (1965). Ph.D. thesis, Bryn Mawr College, Bryn Mawr, Pennsylvania.

Cline, S. G. (1966). *J. Cell. Physiol.* **68,** 157.

Cline, S. G., and Conner, R. L. (1966). *J. Cell. Physiol.* **68,** 149.

Conner, R. L. (1967). *In* "Chemical Zoology" (M. Florkin and B. T. Scheer, eds.), Vol. 1 (G. W. Kidder, ed.), p. 309. Academic Press, New York.

Conner, R. L., and Cline, S. G. (1964). *J. Protozool.* **11,** 484.

Conner, R. L., and Pruett, P. O. (1963). *In* "Progress in Protozoology" (J. Ludvik, J. Lom, and J. Vávra, eds.), p. 143. Academic Press, New York.

Conner, R. L., Goldberg, R., and Kornacker, M. S. (1961a). *J. Gen. Microbiol.* **24,** 239.

Conner, R. L., Goldberg, R., and Kornacker, M. S. (1961b). *J. Gen. Microbiol.* **26,** 437.
Dewey, V. C. (1944). *Arch. Biochem.* **8,** 293.
Dewey, V. C., and Kidder, G. W. (1953a). *Fed. Proc. Fed. Amer. Soc. Exp. Biol.* **12,** 196.
Dewey, V. C., and Kidder, G. W. (1953b). *J. Gen. Microbiol.* **9,** 445.
Dewey, V. C., and Kidder, G. W. (1954). *Proc. Soc. Exp. Biol. Med.* **87,** 198.
Dewey, V. C., and Kidder, G. W. (1964). *Fed. Proc. Fed. Amer. Soc. Exp. Biol.* **23,** 376.
Dunham, P. B. (1964). *Biol. Bull.* **126,** 373.
Dunham, P. B., and Child, F. M. (1961). *Biol. Bull.* **121,** 129.
Elliott, A. M. (1949). *Anat. Rec.* **105,** 47.
Elliott, A. M., and Clark, G. M. (1958a). *J. Protozool.* **5,** 235.
Elliott, A. M., and Clark, G. M. (1958b). *J. Protozool.* **5,** 240.
Elliott, A. M., and Clark, G. M. (1963). *J. Protozool.* **10,** Suppl. 6.
Elliott, A. M., Brownell, L. E., and Gross, J. A. (1954). *J. Protozool.* **1,** 193.
Elliott, A. M., Addison, M. A., and Carey, S. E. (1962). *J. Protozool.* **9,** 135.
Elliott, A. M., Studier, M. A., and Work, J. A. (1964). *J. Protozool.* **11,** 370.
Erickson, E. B., and Conner, R. L. (1968). *J. Protozool.* **15,** Suppl. 13.
Erwin, J. A. (1960). Ph.D. thesis, Syracuse University, Syracuse, New York.
Erwin, J. A. (1962). *J. Protozool.* **9,** 211.
Frank, O., Baker, H., Ziffer, H., Aaronson, S., Hutner, S. H., and Leevy, C. M. (1963). *Science* **139,** 110.
Green, J., Price, S. A., and Gare, L. (1959). *Nature (London)* **184,** 1339.
Guroff, G., and Rhoads, C. A. (1969). *J. Biol. Chem.* **244,** 142.
Hall, R. P. (1944). *Physiol. Zool.* **17,** 200.
Hall, R. P. (1954). *J. Protozool.* **1,** 74.
Hamburger, K., and Zeuthen, E. (1960). *C. R. Trav. Lab. Carlsberg* **32,** 1.
Hoffmann, E. K., and Kramhøft, B. (1969). *Exp. Cell Res.* **56,** 265.
Holm, B. J. (1970). *J. Protozool.* **17,** 615.
Holz, G. G., Jr. (1964). *In* "Biochemistry and Physiology of Protozoa" (S. H. Hutner, ed.), Vol. 3, p. 199. Academic Press, New York.
Holz, G. G., Jr., Erwin, J. A., and Wagner, B. (1961b). *J. Protozool.* **8,** 297.
Holz, G. G., Jr., Wagner, B., Erwin, J., Britt, J. J., and Bloch, K. (1961a). *Comp. Biochem. Physiol.* **2,** 202.
Holz, G. G., Jr., Erwin, J., Wagner, B., and Rosenbaum, N. (1962). *J. Protozool.* **9,** 359.
Hutner, S. H., Nathan, H. A., Baker, H., Sobotka, H., and Aaronson, S. (1959). *Amer. J. Clin. Nutr.* **7,** 407.
Kessler, D. (1961). M.S. thesis, Syracuse University, Syracuse, New York.
Kidder, G. W. (1946). *Arch. Biochem.* **9,** 51.
Kidder, G. W., and Dewey, V. C. (1942). *Growth* **6,** 405.
Kidder, G. W., and Dewey, V. C. (1944). *Biol. Bull.* **87,** 121.
Kidder, G. W., and Dewey, V. C. (1945). *Biol. Bull.* **89,** 131.
Kidder, G. W., and Dewey, V. C. (1947). *Proc. Nat. Acad. Sci. U.S.* **33,** 95.
Kidder, G. W., and Dewey, V. C. (1949a). *Arch. Biochem.* **21,** 66.
Kidder, G. W., and Dewey, V. C. (1949b). *Arch. Biochem.* **21,** 58.
Kidder, G. W., and Dewey, V. C. (1951). *In* "Biochemistry and Physiology of Protozoa" (A. Lwoff, ed.), Vol. 1, p. 323. Academic Press, New York.
Kidder, G. W., and Dewey, V. C. (1961). *Biochem. Biophys. Res. Commun.* **5,** 324.
Kidder, G. W., and Dewey, V. C. (1968). *J. Biol. Chem.* **243,** 826.
Kidder, G. W., Dewey, V. C., Andrews, M. B., and Kidder, R. R. (1949). *J. Nutr.* **37,** 521.

Kidder, G. W., Dewey, V. C., and Parks, R. E., Jr. (1950). *Arch. Biochem.* **27,** 463.

Kidder, G. W., Dewey, V. C., and Parks, R. E., Jr. (1951a). *Proc. Soc. Exp. Biol. Med.* **78,** 88.

Kidder, G. W., Dewey, V. C., and Parks, R. E., Jr. (1951b). *Physiol. Zool.* **24,** 69.

Kramhøft, B. (1970). *C. R. Trav. Lab. Carlsberg* **37,** 343.

Krezanowski, J. Z. (1961). *J. Amer. Pharm. Assoc.* **50,** 421.

Leboy, P. S., Cline, S. G., and Conner, R. L. (1964). *J. Protozool.* **11,** 217.

Lee, D. (1969). *J. Cell. Physiol.* **74,** 295.

Loefer, J. B. (1939). *Physiol. Zool.* **12,** 161.

Lwoff, A., and Lwoff, M. (1937). *C. R. Acad. Sci. Ser.* **D126,** 644.

Munk, N., and Rosenberg, H. (1969). *Biochim. Biophys. Acta* **177,** 629.

Patterson, E. L., Brockman, J. A., Day, F. P., Pierce, J. V., Macchi, M. E., Hoffmann, C. E., Fong, C. T. O., Stokstad, E. L. R., and Jukes, T. H. (1951). *J. Amer. Chem. Soc.* **73,** 5919.

Phelps, A. (1959). *Ecology* **40,** 512.

Prescott, D. M., and Zeuthen, E. (1953). *Acta Physiol. Scand.* **28,** 77.

Pruett, P. O. (1965). Ph.D. thesis, Bryn Mawr College, Bryn Mawr, Pennsylvania.

Pruett, P. O., Conner, R. L., and Pruett, J. R. (1967). *J. Cell. Physiol.* **70,** 217.

Reed, L. J. (1966). *Compr. Biochem.* **14,** 99.

Reed, L. J., DeBusk, B. G., Gunsalus, I. C., and Hornberger, C. S. (1951). *Science* **114,** 93.

Rogers, C. G. (1968). *Can. J. Biochem.* **46,** 331.

Rosenberg, H., and Munk, N. (1969). *Biochem. Biophys. Acta* **184,** 191.

Roth, J. S. (1956). *Exp. Cell Res.* **10,** 146.

Seaman, G. R. (1952). *Proc. Soc. Exp. Biol. Med.* **80,** 308.

Seaman, G. R. (1953). *Physiol. Zool.* **26,** 22.

Seaman, G. R., and Naschke, M. D. (1955). *J. Biol. Chem.* **213,** 705.

Shenoy, K. G., and McLaughlan, J. M. (1959). *Can. J. Biochem. Physiol.* **37,** 1388.

Shug, A. L., Elson, C., and Shrago, E. (1969). *J. Nutr.* **99,** 379.

Slater, J. V. (1952). *Physiol. Zool.* **25,** 283.

Slater, J. V. (1957). *Biol. Bull.* **112,** 390.

Slater, J. V., and Tremor, J. W. (1962). *Biol. Bull.* **122,** 298.

Snell, E. E., and Broquist, H. P. (1949). *Arch. Biochem.* **23,** 326.

Stokstad, E. L. R., Hoffmann, C. E., Regan, M. A., Fordham, D., and Jukes, T. H. (1949). *Arch. Biochem.* **20,** 75.

Tittler, I. A., Belsky, M. M., and Hutner, S. H. (1952). *J. Gen. Microbiol.* **6,** 85.

Wykes, J. R., and Prescott, D. M. (1968). *J. Cell. Physiol.* **72,** 173.

Zeuthen, E. (1968). *J. Protozool.* **15,** Suppl. 36.

CHAPTER 9

Effects of Radiation, Drugs, and Hydrostatic Pressure

Radiation

Tetrahymena pyriformis grows well in the absence of light (Gross, 1962); and intense visible light kills those organisms that have been grown in darkness. The cells accumulate, in the dark, a red pigment which contains free protoporphyrin and protoporphyrin combined with a lipid constituent and which probably acts as a photodynamic sensitizer to produce the lethal effect (Rudzinska and Granick, 1953). An indirect inhibition of growth by strong visible light is caused by destruction of folate, thiamine, and riboflavin in the medium (Dewey and Kidder, 1953; Phelps, 1959; Lee, 1969). Synchronization of cell division of *T. pyriformis* can be achieved by a sudden increase or decrease in irradiance occurring after a critical time late in the exponential (ultradian) growth phase (Wille and Ehret, 1968).

The ciliate is quite resistant to ultraviolet radiation, with an LD_{50} of 1250 krad (Calkins, 1964). The presence of methionine in the medium increases survival (Sullivan, 1959a,b). Further, cells are protected from the deleterious effect of transfer to an irradiated medium by growth in a medium containing large amounts of cysteine (Sullivan *et al.*, 1962). Organisms after heat treatment are more resistant to ultraviolet injury, probably due to their accumulation of large amounts of deoxyribonucleic acid (Iverson and Giese, 1957). Starvation lowers resistance to ultraviolet light, but exposure of the cells to nutrients does not immediately increase resistance. Apparently the nutrients must be incorporated before they are effective (Giese *et al.*, 1954).

Ultraviolet light delays division of cells synchronized by heat, but the substance that is sensitive to ultraviolet light is different from that which is sensitive to heat (Nachtwey and Giese, 1968). In UV-irradiated cells, there is also a decline in respiration (Giese, 1942), a loss of intracellular potassium ions (Andrus, 1961), an increase in succinate dehydrogenase activity at the end of heat treatment and during the first synchronous division (Sullivan and Sparks, 1961), and an increase in isocitrate dehydrogenase activity (Sullivan and Ehrman, 1966).

An apparent dispute exists regarding deoxyribonucleic acid synthesis in UV-irradiated cells. One report states that after exposure of the cells, both macronuclear and micronuclear synthesis proceeds at lower rates; and a smaller total amount of each is produced (Harrington, 1960). Further evidence for a decrease in synthesis is a drop in the incorporation of precursors into deoxyribonucleic acid (Iverson and Giese, 1957). However, another report shows that individual cells of *T. pyriformis* which are irradiated with ultraviolet light accumulate 3 times the amount of deoxyribonucleic acid as do normal cells prior to division (Shepard, 1965). During the next two divisions of the treated cells, the usual amount is synthesized; and the excess is extruded in large and numerous bodies so that after three divisions, the normal complement is present.

The ionizing properties of a beam of high-energy electrons causes a nonspecific, but covalent, binding of amino acids and carbohydrates to nucleic acids of *T. pyriformis* (Byfield *et al.*, 1970). The reactions may be similar to those linking deoxyribonucleic acid and protein following ultraviolet irradiation.

Tetrahymena are also highly resistant to X-radiation. The LD_{50} varies with the strain used; the values range from 120 to 150 krad (Wells, 1960; Roth and Buccino, 1963; Calkins, 1964). Allylisopropylacetamide sensitizes the cells to X-rays (Roth and Buccino, 1963). Sublethal doses of X-radiation block cell replication shortly after the beginning of exposure (Schmid, 1967), but cell growth in volume continues during irradiation, demonstrating that such growth is independent of the division cycle.

Following X-irradiation, there is a lag phase before multiplication is resumed by either cells in exponential growth or cells undergoing synchronous division (Roth, 1962; Ducoff, 1956). The presence of metabolites during irradiation does not affect the lag period or the subsequent rate of division (Roth, 1962). Since cells in synchrony have much more deoxyribonucleic acid, the division delay is probably not due to inhibition of synthesis of this material. Deoxyribonucleic acid synthesis ceases early in the irradiation period but then begins anew during the exposure at a time when cell division is still blocked (Schmid, 1967). This late synthesis may represent the start of repair.

There is an unconfirmed report that X-radiation can induce mutations in *T. pyriformis*. Some of the progeny of an irradiated strain which can make serine do not have this ability (Elliott and Clark, 1957). High doses of X-radiation produce chromosomal aberrations and other nuclear abnormalities (Elliott *et al.*, 1954; Wells, 1962), and cells suffering such damage usually fail to give rise to viable progeny. When heavily irradiated cells are mated with normal cells, some of the progeny are haploid, owing to the destruction of the genetic apparatus in the irradiated cell (Elliott and Clark, 1956).

Other effects of X-rays are that exposed cells have decreased fumarase activity (Sullivan and Snyder, 1962) and increased isocitrate dehydrogenase activity (Sullivan and Ehrman, 1966). There is also decreased respiration if the cells are suspended in water rather than in phosphate buffer (Roth, 1962; Roth and Eichel, 1955). X-Radiation prevents the decrease in succinate dehydrogenase which occurs in heat-treated cells at the end of heat treatment and during the first synchronous division (Sullivan and Boyle, 1961). The ribonuclease and deoxyribonuclease of intact cells are not changed significantly, but irradiation of homogenates decreases the activity of these enzymes to 50% of normal (Eichel and Roth, 1953). Also in the homogenates, the oxidation of lactate and β-hydroxybutyrate is reduced (Roth, 1962); and there is appreciable destruction of succinate dehydrogenase, glutamate dehydrogenase, malate dehydrogenase, and catalase (Roth and Eichel, 1955). The catalase activity recovers when the homogenate is allowed to stand at 0°C.

Catalase is inactivated by γ-irradiation, which also decreases cell permeability. Sulfhydryl compounds, when added to *T. pyriformis* cells in buffer, protect against γ-rays (Van de Vijver, 1969a,b).

Repair of Radiation Damage

Investigation of four strains of *Tetrahymena pyriformis* shows that nonsemiconservative synthesis of deoxyribonucleic acid occurs in the dark after exposure of the cells to UV light or X-rays (Brunk and Hanawalt, 1967). This synthesis, involving repair of radiation damage, follows excision of the defect by nucleases. Bacteria have a similar mechanism of repair (Pettijohn and Hanawalt, 1964). In accomplishing repair, the activity of one form of *T. pyriformis* deoxyribonucleic acid polymerase increases (Keiding and Westergaard, 1969). The activity increased is probably active only in restoring damaged strands and is not involved in replication. The average length of restored regions in deoxyribonucleic acid is about 5000 nucleotides. This process is completed before normal replication occurs (Brunk and Hanawalt, 1969). Attempts to demonstrate in *T. pyriformis* the repair enzymes found in extracts of other organisms have been unsuccessful, probably due to the presence of very active deoxyribonucleases (Schroder *et al.*, 1969).

The faulty deoxyribonucleic acid produced by ultraviolet irradiation contains thymine dimers, which can be isolated and identified after acid hydrolysis (Whitson *et al.*, 1968). Immediately following irradiation of cells, the dimers are found exclusively in the trichloroacetic acid-insoluble material; but later they appear in the soluble portion. They gradually disappear from the cell; after 48 hr, only a small portion can be found. Monomerization is thought to occur by an enzymatic process similar to that in other systems (Francis and Whitson, 1969).

Cells of *T. pyriformis* show a cyclic variation in sensitivity to UV light (Sullivan, 1959b; Calkins, 1968) but not to X-rays (Calkins, 1968), and caffeine-treated cells endure higher doses of UV light better than lower doses (Calkins, 1968). To explain these phenomena, Calkins (1967, 1968) shows that there are two basic repair mechanisms—a normal, caffeine-sensitive system and a more efficient one, which is less sensitive to caffeine and which is active only after radiation damage. The induced, caffeine-insensitive system accounts for the resistance to high doses. For UV-induced lesions, it has a greater effect than the other system throughout most of the cell cycle. However, the caffeine-sensitive system is unusually efficient for repair during the G_2 phase, which accounts for the variation in the sensitivity to UV light at different stages of cell growth.

Ultraviolet injury can be reversed by visible or near UV light (Christenssen and Giese, 1956; Iverson and Giese, 1957; Brunk and Hanawalt, 1967). A small percentage of cells that have been exposed to a lethal dose of UV radiation can divide one or two times afterward (Calkins, 1962), and photoreversal can be effective even after the cell has divided. Photoreactivation cleaves the thymine-containing dimers (Francis and Whitson, 1969) and reduces the amount of repair (Brunk and Hanawalt, 1967). The maxima for photoreversal are at 365 and 436 nm, and the rate is temperature-dependent. The growth delay of cells after ultraviolet exposure can be shortened from 20 to 2 hr by an amount of visible light which reduces the number of dimers only slightly. Further, the same number of dimers is formed by equal exposure to light of 265 and 280 nm; but exposure at 280 nm kills twice as many cells (Whitson, 1969). One may conclude that, although the formation of dimers is an outstanding characteristic of UV irradiation, their presence is not responsible for all of the lethal effect. Another, yet unidentified, lesion is indicated. Photoreactivation is not observed for X-ray damage.

The Photodynamic Effect

A number of compounds serve as photosensitizers in suspensions of *Tetrahymena.* The photodynamic activity is an oxygen-dependent phenomenon in which a combination of light energy and chemical sensitizer produces toxic effects induced by neither one alone. The procedure, first developed for detecting polycyclic hydrocarbons (Hull, 1962), may be valuable in distinguishing those which are carcinogenic (Epstein *et al.*, 1963). A study of cell uptake and photodynamic activity of twenty-five polycyclic compounds shows that uptake varies inversely with molecular size and that a high intracellular concentration is prerequisite, but not sufficient, for high photodynamic activity (Small *et al.*, 1967). The presence of Tween in the medium protects against the photodynamic toxicity of benzo(*a*)pyrene (Epstein and Niskanen, 1967). The protection is due to elution by Tween of the polycyclic hydrocarbon from the mitochondria.

Compounds that reverse the photodynamic toxicity of benzo(*a*)pyrene and other sensitizers are called antioxidants, and a number of these are known. The most potent are certain tocopherols, certain aminoazobenzenes, lecithin, and tetraethylthiuramdisulfide (Antabuse) (Epstein *et al.*, 1965, 1966; Epstein and Saporoschetz, 1966). In general, antioxidants yield similar responses in the photodynamic bioassay regardless of which of the photosensitizing agents is used (Epstein *et al.*, 1965).

Tests for toxicity to *T. pyriformis* of ninety-seven antioxidants reveal that the cytotoxicity is not associated with antioxidant potency as measured by the photodynamic assay. However, cytotoxicity is generally associated with oral cytotoxicity to rats (Epstein *et al.*, 1967).

Tetrahymena as a Toxicological Tool

The tests for toxicity of the antioxidants just mentioned are examples of experiments involving many varieties of toxic compounds. *Tetrahymena* provide a simple, relatively inexpensive method for evaluating inhibitory substances. Of course, inhibition of its growth does not always correlate with toxicity to other organisms. The value and practicality of using microorganisms in screening antitumor agents has been considered (Hutner *et al.*, 1958; Hutner, 1964), and a convenient technique for growing *T. pyriformis* and certain other protozoans in agar for testing of such agents is available (West *et al.*, 1962).

A number of studies have been made to determine the utility of *T. pyriformis* in detecting antitumor agents. Of sixteen microbiological assay organisms, *T. pyriformis* was most effective in detecting antitumor compounds (Foley, 1958a,b). In other studies (Johnson *et al.*, 1962; Price *et al.*, 1962), it was almost as effective as cultured mammalian tumor (HeLa) cells in detecting the antitumor activity of known anticancer agents. However, the organism was of little use as a detector of such compounds in fermentation beers, perhaps due to the low concentration of the toxic substances.

The biological effects of the carcinogen and anticarcinogen, 4-nitroquinoline 1-oxide, on *T. pyriformis* are known in some detail. The compound inhibits cell division, produces nuclear inclusions, and causes distorted karyokinesis, with uneven distribution of deoxyribonucleic acid and irregularity in cell sizes. A clone with atypical multiplication has been isolated from cultures treated with 4-nitroquinoline 1-oxide (Mita *et al.*, 1965, 1966). Tryptophan, 5-methyltryptophan, and tryptamine competitively annul the toxicity (Zahalsky *et al.*, 1962). Nine carcinogenic derivatives of 4-nitroquinoline 1-oxide induce nuclear derangement in synchronized cells, but nine noncarcinogenic derivatives do not (Mita *et al.*, 1969). 4-Nitroquinoline 1-oxide binds strongly to cytoplasmic proteins, but the noncarcinogenic 4-aminoquinoline 1-oxide does not attach (Mita *et al.*, 1970).

The carcinogenic compounds, 2-aminofluorene, α-naphthylene, and β-naphthylene, are toxic to *T. pyriformis* (Roth, 1954). They reduce respiration, but this effect is partially reversed by pantothenate and completely reversed by a liver extract. They also inhibit succinate oxidase activity, but ribonuclease and deoxyribonuclease are not affected. Aminoazobenzene dyes, some of which are carcinogenic, are toxic and cause the organism to assume irregular shapes (Jacob, 1958).

Some antimalarial agents inhibit growth of *T. pyriformis*. Quinacrine is more effective than proguanil, primaquine, and quinine (Clancey, 1968). It may have some effect on nucleic acid synthesis (Chou and Ramanathan, 1968; Chou *et al.*, 1968). Quinine inhibits ribonucleic acid, deoxyribonucleic acid, lipid, and protein synthesis, perhaps by blocking metabolism of an energy source (Conklin *et al.*, 1969a,b).

The phenothiazine tranquilizer, chlorpromazine, immobilizes *T. pyriformis* and increases its permeability (Nathan and Friedman, 1962). Test of a series of phenothiazines shows that the alteration in permeability is independent of the effectiveness of immobilization (Guttman and Friedman, 1963a). The immobilization by chlorpromazine is reversed by lecithin, Tween 80, and calcium ions; a number of sterols are not effective (Guttman and Friedman, 1963b). The lethal effect of chlorpromazine is antagonized by cysteine and chlordiazepoxide (Forrest *et al.*, 1964). Phenothiazines inhibit glucose uptake and reduce the amount of lipid, ribonucleic acid, and deoxyribonucleic acid in the cell (Rogers, 1966).

The antihistamines, diphenylhydramine, tripelennamide, and pheniramine, inhibit the growth and motility of *T. pyriformis* (Sanders and Nathan, 1959). Ornithine, proline, and histidine reverse the effect on growth; and histidine reverses the effect on motility.

The antimitotic agents, Colcemid and colchicine, block fission of *Tetrahymena limacis* and *T. pyriformis* (Dysart and Corliss, 1960; Wunderlich and Peyk, 1969a,b,c). Both of these agents produce nuclear disruption and cause chromatin aggregation in the nucleoplasm. *Tetrahymena pyriformis* is able to recover from the effects of low levels of Colcemid and colchicine; and, after recovery, the subsequent divisions are synchronized (Wunderlich and Peyk, 1969a,b,c). In heat-synchronized cells, low concentrations of Colcemid or colchicine cause division delays and inhibit nucleic acid, but not protein, synthesis (Kuzmich and Zimmerman, 1970; Nelsen, 1970). Colchicine also blocks the development of oral structures (Nelsen, 1970).

Many of the benzimidazoles, benzotriazoles, and quinoxalines which contain a nitro group on the benzene ring are inhibitory to *T. pyriformis* (Greer, 1958). Other toxic agents are aureomycin (Gross, 1955), 2-mercaptoethanol (Gavin and Frankel, 1966; Mazia and Zeuthen, 1966), and some radiopaque agents used in angiology (Mark *et al.*, 1963).

Inhibitory compounds with specific effects on protein and nucleic acid synthesis are considered in Chapters 5 and 6, respectively.

Hydrostatic Pressure

Tetrahymena pyriformis is resistant to high hydrostatic pressure. No effect is noted at a pressure of 100 atm, but division is halted if the pressure is increased to 250 atm. Maintaining such a force causes cleavage furrows to be resorbed; but division is resumed after decompression (Macdonald, 1967a,b). Pulse treatments of very high pressure can cause division delays and blockage of oral differentiation in heat-treated cells (Lowe-Jinde and Zimmerman, 1969; Simpson and Williams, 1970). A pressure of 544 atm causes oxygen consumption by the organism to drop to 46% of normal (Macdonald, 1965). At 350 atm, ribonucleic acid synthesis drops 50% (Yuyama and Zimmerman, 1969), and cells migrate to the anode in an electrical field, whereas normally they migrate to the cathode (Zimmerman and Murakami, 1968). Treatment for 2 min at 350 atm reduces the number of polysomes present in heat-treated cells (Hermolin and Zimmerman, 1969), but microsomes from pressurized cells incorporate phenylalanine as efficiently as control microsomes (Letts and Zimmerman, 1970). Since hydrostatic pressure inhibits protein and ribonucleic acid synthesis, it may exert its primary effect by preventing the accumulation of proteins necessary for division (Lowe-Jinde and Zimmerman, 1971).

REFERENCES

Andrus, W. D. (1961). Ph.D. thesis, Stanford University, Palo Alto, California.

Brunk, C. F., and Hanawalt, P. C. (1967). *Science* **158,** 663.

Brunk, C. F., and Hanawalt, P. C. (1969). *Radiat. Res.* **38,** 285.

Byfield, J. E., Lee, Y. C., and Bennett, L. R. (1970). *Nature (London)* **225,** 859.

Calkins, J. (1962). *Nature (London)* **196,** 686.

Calkins, J. (1964). *Photochem. Photobiol.* **3,** 143.

Calkins, J. (1967). *Int. J. Radiat. Biol.* **13,** 283.

Calkins, J. (1968). *Photochem. Photobiol.* **8,** 115.

Chou, S. C., and Ramanathan, S. (1968). *Life Sci.* **7,** 1053.

Chou, S. C., Ramanathan, S., and Cutting, W. C. (1968). *Pharmacology* **1,** 60.

Christenssen, E., and Giese, A. C. (1956). *J. Gen. Physiol.* **38,** 513.

Clancy, C. F. (1968). *Amer. J. Trop. Med. Hyg.* **17,** 359.

Conklin, K. A., Chou, S. C., and Ramanathan, S. (1969a). *Fed. Proc. Fed. Amer. Soc. Exp. Biol.* **28,** 361.

Conklin, K. A., Chou, S. C., and Ramanathan, S. (1969b). *Pharmacology* **2,** 247.

Dewey, V. C., and Kidder, G. W. (1953). *J. Gen. Microbiol.* **9,** 445.

Ducoff, H. S. (1956). *Exp. Cell Res.* **11,** 218.

Dysart, M. P., and Corliss, J. O. (1960). *J. Protozool.* **7,** Suppl. 10.

Eichel, H. J., and Roth, J. S. (1953). *Biol. Bull.* **104,** 351.

Elliott, A. M., and Clark, G. M. (1956). *J. Protozool.* **3,** 181.

Elliott, A. M., and Clark, G. M. (1957). *Biol. Bull.* **113,** 345.

Elliott, A. M., Hayes, R. E., and Byrd, J. R. (1954). *Biol. Bull.* **107,** 309.

Epstein, S. S., and Niskanen, E. E. (1967). *Exp. Cell Res.* **46,** 211.

Epstein, S. S., and Saporoschetz, I. B. (1966). *Life Sci.* **5,** 783.

Epstein, S. S., Burroughs, M., and Small, M. (1963). *Cancer Res.* **23,** 35.

Epstein, S. S., Saporoschetz, I. B., Small, M., Park, W., and Mantel, N. (1965). *Nature (London)* **208,** 655.

Epstein, S. S., Forsyth, J., Saporoschetz, I. B., and Mantel, N. (1966). *Radiat. Res.* **28,** 322.

Epstein, S. S., Saporoschetz, I. B., and Hutner, S. H. (1967). *J. Protozool.* **14,** 238.

Foley, G. E., McCarthy, R. E., Binns, V. M., Snell, E. E., Guirard, B. M., Kidder, G. W., Dewey, V. C., and Thayer, P. S. (1958a). *Ann. N.Y. Acad. Sci.* **76,** 413.

Foley, G. E., Eagle, H., Snell, E. E., Kidder, G. W., and Thayer, P. S. (1958b). *Ann. N.Y. Acad. Sci.* **76,** 952.
Forrest, I. S., Quesada, F., and Dietshman, G. L. (1964). *Proc. West. Pharmacol. Soc.* **7,** 42.
Francis, A. A., and Whitson, G. L. (1969). *Biochim. Biophys. Acta* **179,** 253.
Gavin, R. H., and Frankel, J. (1966). *J. Exp. Zool.* **161,** 63.
Giese, A. (1942). *J. Cell Comp. Physiol.* **20,** 35.
Giese, A., Brandt, C. L., Jacobson, C., Shepard, D. C., and Sanders, R. T. (1954). *Physiol. Zool.* **27,** 71.
Greer, S. B. (1958). *J. Gen. Microbiol.* **18,** 543.
Gross, J. A. (1955). *J. Protozool.* **2,** 242.
Gross, J. A. (1962). *J. Protozool.* **9,** 415.
Guttman, H. N., and Friedman, W. (1963a). *Fed. Proc. Fed. Amer. Soc. Exp. Biol.* **22,** 569.
Guttman, H. N., and Friedman, W. (1963b). *Trans. N.Y. Acad. Sci.* **26,** 75.
Harrington, J. D. (1960). Ph.D. thesis, Catholic University of America, Washington, D. C.
Hermolin, J., and Zimmerman, A. M. (1969). *Cytobios* **1,** 247.
Hull, R. W. (1962). *J. Protozool.* **9,** Suppl. 18.
Hutner, S. H. (1964). *J. Protozool.* **11,** 1.
Hutner, S. H., Nathan, A., Aaronson, S., Baker, H., and Scher, S. (1958). *Ann. N.Y. Acad. Sci.* **76,** 457.
Iverson, R. M., and Giese, A. C. (1957). *Exp. Cell Res.* **13,** 213.
Jacob, M. I. (1958). *Diss. Abstr.* **18,** 368.
Johnson, I. S., Simpson, P. J., and Cline, J. C. (1962). *Cancer Res.* **22,** 617.
Keiding, J., and Westergaard, O. (1969). *J. Protozool.* **16,** Suppl. 33.
Lee, D. (1969). *J. Cell. Physiol.* **74,** 295.
Letts, P. J., and Zimmerman, A. M. (1970). *J. Protozool.* **17,** 593.
Lowe-Jinde, L., and Zimmerman, A. M. (1969). *J. Protozool.* **16,** 226.
Lowe-Jinde, L., and Zimmerman, A. M. (1971). *J. Protozool.* **18,** 20.
Kuzmich, M. J., and Zimmerman, A. M. (1970). *Fed. Proc. Fed. Amer. Soc. Exp. Biol.* **29,** 849.
Macdonald, A. G. (1965). *Exp. Cell Res.* **40,** 78.
Macdonald, A. G. (1967a). *Exp. Cell Res.* **47,** 569.
Macdonald, A. G. (1967b). *J. Cell. Physiol.* **70,** 127.
Mark, M. F., Imparato, A. M., Hutner, S. H., and Baker, H. (1963). *Angiology* **14,** 383.
Mazia, D., and Zeuthen, E. (1966). *C. R. Trav. Lab. Carlsberg* **35,** 341.
Mita, T., Tokuzen, R., Fukuoka, F., and Nakahara, W. (1965). *Gann* **56,** 293.
Mita, T., Tokuzen, R., Fukuoka, F., and Nakahara, W. (1966). *Gann* **57,** 273.
Mita, T., Kawazoe, Y., and Araki, M. (1969). *Gann* **60,** 155.
Mita, T., Munakata, H., and Nakahara, W. (1970). *Exp. Cell Res.* **60,** 299.
Nachtwey, D. S., and Giese, A. C. (1968). *Exp. Cell Res.* **50,** 167.
Nathan, H. A., and Friedman, W. (1962). *Science* **135,** 793.
Nelsen, E. M. (1970). *J. Exp. Zool.* **175,** 69.
Pettijohn, D. E., and Hanawalt, P. C. (1964). *J. Mol. Biol.* **9,** 395.
Phelps, A. (1959). *Ecology* **40,** 512.
Price, K. E., Buck, R. E., Schlein, A., and Siminoff, P. (1962). *Cancer Res.* **22,** 885.
Rogers, C. G. (1966). *Can. J. Biochem.* **44,** 1493.
Roth, J. S. (1954). *Cancer Res.* **14,** 346.

Roth, J. S. (1962). *J. Protozool.* **9,** 142.
Roth, J. S., and Buccino, G. (1963). *Radiat. Res.* **19,** 193.
Roth, J. S., and Eichel, H. J. (1955). *Biol. Bull.* **108,** 308.
Rudzinska, M. A., and Granick, S. (1953). *Proc. Soc. Exp. Biol. Med.* **83,** 525.
Sanders, M., and Nathan, H. A. (1959). *J. Gen. Microbiol.* **21,** 264.
Schmid, P. (1967). *Biochem. Biophys. Res. Commun.* **26,** 615.
Schroder, E. W., Meskill, V. P., and Garner, J. G. (1969). *J. Protozool.* **16,** Suppl. 11.
Shepard, D. C. (1965). *Exp. Cell Res.* **38,** 570.
Simpson, R. E., and Williams, N. E. (1970). *J. Exp. Zool.* **175,** 85.
Small, M., Mantel, N., and Epstein, S. S. (1967). *Exp. Cell Res.* **45,** 206.
Sullivan, W. D. (1959a). *Trans. Amer. Microsc. Soc.* **78,** 181.
Sullivan, W. D. (1959b). *Trans. Amer. Microsc. Soc.* **78,** 285.
Sullivan, W. D., and Boyle, J. V. (1961). *Broteria Ser. Cienc. Natur.* **30,** 77.
Sullivan, W. D., and Ehrman, R. A. (1966). *Broteria Ser. Cienc. Natur.* **35,** 93.
Sullivan, W. D., and Snyder, R. L. (1962). *Exp. Cell Res.* **28,** 239.
Sullivan, W. D., and Sparks, J. T. (1961). *Exp. Cell Res.* **23,** 536.
Sullivan, W. D., McCormick, S. J., and McCormick, E. A. (1962). *Trans. Amer. Microsc. Soc.* **81,** 80.
Van de Vijver, G. (1969a). *Enzymologia* **36,** 371.
Van de Vijver, G. (1969b). *Enzymologia* **36,** 375.
Wells, C. (1960). *J. Cell. Comp. Physiol.* **55,** 207.
Wells, M. M. (1962). *Diss. Abstr.* **22,** 3824.
West, R. A., Jr., Barbera, P. W., Kolar, J. H., and Murrell, C. B. (1962). *J. Protozool.* **9,** 65.
Whitson, G. L. (1969). *J. Cell Biol.* **43,** 157A.
Whitson, G. L., Francis, A. A., and Carrier, W. L. (1968). *Biochim. Biophys. Acta* **161,** 285.
Wille, J. J., Jr., and Ehret, C. F. (1968). *J. Protozool.* **15,** 785.
Wunderlich, F., and Peyk, D. (1969a). *Naturwissenschaften* **56,** 285.
Wunderlich, F., and Peyk, D. (1969b). *Exp. Cell Res.* **57,** 142.
Wunderlich, F., and Peyk, D. (1969c). *Experientia* **25,** 1278.
Yuyama, S., and Zimmerman, M. (1969). *J. Cell Biol.* **43,** 162A.
Zahalsky, A. C., Keane, M., Hutner, S. H., Kittrell, M., and Amsterdam, D. (1962). *J. Protozool.* **9,** Suppl. 12.
Zimmerman, A. M., and Murakami, T. H. (1968). *J. Cell Biol.* **39,** 147A.

CHAPTER 10

Evolution

Phylogenetic Relationships

Biologists are fond of drawing phylogenetic trees, which show the evolutionary relationship of all living things. Those who do this are hard pressed to find a place in the scheme for Protozoa, due mainly to the lack of a fossil record and the lack of ontogeny to recapitulate phylogeny. The relationship of these organisms to bacteria, fungi, higher plants, and higher animals must at present be deduced from meager and often conflicting data. Some scant biochemical evidence is found in the pecularities of biosynthesis of fatty acids by *Tetrahymena pyriformis* (Erwin and Bloch, 1963). The ciliate has a pathway intermediate between that of flagellates and vertebrates, supporting the theory that Metazoa arose from flagellates by way of ciliates (Hanson, 1958).

From the information known about *Tetrahymena,* a disputed point is whether these organisms are plants or animals (Holz, 1966). There are strong arguments in favor of calling them animals. When viewed under the microscope, they are seen to be motile, to ingest food, and to have no choloroplasts. They have the same amino acid requirements as man and the rat (Kidder and Dewey, 1951), and the vitamin requirements for all three are quite similar. *Tetrahymena* contains, as do higher animals, glycogen as a storage form of carbohydrate (Ryley, 1952); it breaks down the glycogen anaerobically to lactate (Warnock and van Eys, 1962). The glycogen is formed from uridine diphosphate (UDP)-glucose (Cook *et al.*, 1968), a precursor of glycogen in animals but not in plants. The biosynthesis of phosphatidyl serine proceeds in a manner similar to that in animals, but not to that in bacteria (Dennis and Kennedy, 1970); and the ribonucleic acid polymerase is similar to that in animals, but not to that in bacteria (Byfield *et al.*, 1970). The organisms also possess hemoglobin (Keilin and Ryley, 1953) and probably contain *N*-phosphorylarginine as a phosphagen (Robin and Viala, 1966). Neither substance is found in plants.

This formidable evidence is not conclusive, however. *Tetrahymena* require lipoate (Seaman, 1955); animals are able to synthesize this vitamin. The ciliate makes a vitamin B_{12}-like compound (Erwin and Holz, 1962); but animals require vitamin B_{12}. The arginine dihydrolase enzymes, used by *T. pyriformis* to convert arginine to citrulline (Hill and Chambers, 1967), have been found in bacteria and yeast, but not in animals. The ciliate also has a bacterial-type cytochrome c (Yamanaka *et al.*, 1968). Finally, *T. pyriformis*, like plants, has an operative glyoxalate cycle (Hogg and Kornberg, 1963) and contains a pentacyclic triterpenoid (Mallory *et al.*, 1963), both of which have been found only in plants.

One cannot, on the basis of this evidence, describe *Tetrahymena* as either "plant" or "animal." The ciliate has biochemical similarities to both kingdoms but really belongs to neither group. But if one cannot place Protozoa precisely in the phylogenetic scheme, one can show that members of the phylum are related. Hybridization studies with the deoxyribonucleic acids of a number of protozoans show that all have nucleotide sequences in common with *Paramecium aurelia* (Gibson, 1966). The very high degree of hybridization between the deoxyribonucleic acid of *Paramecium* and *Tetrahymena* implies that the two cilates are closely akin.

Evolutionary Oddities

The biochemistry of *Tetrahymena* is unusual in some respects, but unfortunately the pecularities allow no definite conclusions of an evolutionary nature to be made. The complete absence of the urea cycle enzymes is one such eccentricity (Dewey *et al.*, 1957; Hill and Chambers, 1967). These enzymes serve different organisms in different ways, and most living things have at least a portion of the cycle. Cohen and Brown (1960), reviewing the literature on the urea cycle, note that some (usually not all) of the enzymes are present in bacteria, yeast, algae, higher plants, primitive fish, amphibians, reptiles, and mammals. With the possible exception of modern fish (Osteichthes), for which only two of the urea cycle enzymes have been assayed, *Tetrahymena* appear to stand alone in lacking the entire urea cycle.

Carbamylphosphate and the enzymes involved in its synthesis may be important from the standpoint of evolution. Carbamylphosphate is an energy-rich, organic compound which can be formed from the inorganic reactants cyanate and phosphate. It may have been one of the first biochemical compounds. Carbamylphosphate synthetase accomplishes the primary fixation of ammonia and carbon dioxide, utilzing two molecules of adenosine triphosphate (ATP) per molecule of product formed. The enzyme is present in the livers of vertebrates (Grisolia and Cohen, 1953) and is probably present in yeast (Cohen and Brown, 1960) and higher plants (Bone, 1959). Bacteria and yeast possess a similar enzyme, carbamate kinase, which catalyzes the reversible synthesis of carbamylphosphate (Knivett, 1954). In this case, only one molecule of ATP is required in the production of a molecule of the product.

Tetrahymena may lack the ability to form carbamylphosphate by either mechanism. Since the urea cycle is absent, there is no requirement for its synthesis in this respect. The other known pathway in which carbamylphosphate is used is that for the synthesis of pyrimidines. The utilization of carbamylphosphate by *Tetrahymena* for this purpose is avoided because the ciliate requires preformed pyrimidines (Kidder and Dewey, 1951). Furthermore, *Tetrahymena* may not possess carbamate kinase, as the enzyme has been found only in organisms that produce ATP during the conversion of citrulline to ornithine. *Tetrahymena* accomplish this reaction, but no ATP is produced (Hill and Chambers, 1967).

A disturbing fact related to the evolution of *Tetrahymena* is that the organism can make ubiquinones (Miller, 1965), which contain a benzene ring, but cannot make the amino acid, phenylalanine, which also has this structure. *Tetrahymena* make the ubiquinones from shikimate, which, in other microorganisms, is also a precursor of phenylalanine. In an environment devoid of phenylalanine, *Tetrahymena* would perish for lack of the few enzymes which convert shikimate to the amino acid. This arrangement does not appear to be an evolutionary advantage to the cell, for it makes the organism vulnerable in a changing environment.

Loss of Synthetic Ability

It used to be fashionable to say that the most highly evolved organisms were those that had lost much of the synthetic ability possessed by bacteria and other prokaryotic organisms (Van Niel, 1949). It is not entirely clear how evolution could have proceded, at first with the acquisition of new types of macromolecules and new pathways until structures such as bacteria were developed and, then, changed courses with progressive loss of the hard-won capabilities. Without doubt, if one applies to *Tetrahymena* an evolutionary yardstick based upon the loss of synthetic ability, one must conclude that this ciliate is highly evolved. It requires for growth 10 amino acids, 6 vitamins, guanine, and uracil (Kidder and Dewey, 1951). The organism has none of the urea cycle enzymes (Dewey *et al.*, 1957; Hill and Chambers, 1967) and probably can make neither sterols (Williams *et al.*, 1966), glutathione (Hill, 1964), nor carbamylphosphate (Hill, 1964).

However, there is some question as to the rating of organisms on such an evolutionary scale. Is man necessarily more evolved than, for instance, the yeasts? Is *Tetrahymena* reaching the pinnacle of evolutionary success because it cannot make a large number of amino acids and vitamins? Very likely, the losses of enzymes were incidental as evolving organisms developed other structures and processes. One criterion that could be applied as easily as the loss of synthetic ability is the ability to survive under conditions that the organism is likely to encounter. If we apply this criterion, then many will agree that yeasts are more highly evolved than man, for they will persist long after *Homo sapiens* has joined the dinosaurs in the void of extinction. No one knows why the dinosaurs left the earth; but if no pathogenic microbe removes man from the scene, then he is likely to remove himself. He already has the ability; and if he retains such power, it is a mathematical certainty that he will, at some time, use it.

REFERENCES

Bone, D. H. (1959). *Plant Physiol.* **34,** 171.

Byfield, J. E., Lee, Y. C., and Bennett, L. R. (1970). *Biochim. Biophys. Acta* **204,** 610.

Cohen, P. P., and Brown, G. W., Jr. (1960). *In* "Comparative Biochemistry" (M. Florkin and H. S. Mason, eds.), Vol. 2, p. 161. Academic Press, New York.

Cook, D. E., Rangaraj, N. I., Best, N., and Wilken, D. R. (1968). *Arch. Biochem. Biophys.* **127,** 72.

Dennis, E. A., and Kennedy, E. P. (1970). *J. Lipid Res.* **11,** 394.

Dewey, V. C., Heinrich, M. R., and Kidder, G. W. (1957). *J. Protozool.* **4,** 211.

Erwin, J., and Bloch, K. (1963). *J. Biol. Chem.* **238,** 1618.

Erwin, J. A., and Holz, G. G., Jr. (1962). *J. Protozool.* **9,** 211.

Gibson, I. (1966). *J. Protozool.* **13,** 650.

Grisolia, S., and Cohen, P. P. (1953). *J. Biol. Chem.* **204,** 753.

Hanson, C. D. (1958). *Syst. Zool.* **7,** 16.

Hill, D. L. (1964). Ph.D. thesis, Vanderbilt University.

Hill, D. L., and Chambers, P. (1967). *Biochim. Biophys. Acta* **148,** 435.

Hogg, J. F., and Kornberg, H. L. (1963). *Biochem. J.* **86,** 462.

Holz, G. G., Jr. (1966). *J. Protozool.* **13,** 2.

Keilin, D., and Ryley, J. F. (1953). *Nature* (*London*) **172,** 451.
Kidder, G. W., and Dewey, V. C. (1951). *In* "Biochemistry and Physiology of Protozoa" (A. Lwoff, ed.), Vol. 1, p. 323. Academic Press, New York.
Knivett, V. A. (1954). *Biochem. J.* **56,** 606.
Mallory, F. B., Gordon, J. T., and Conner, R. L. (1963). *J. Amer. Chem. Soc.* **85,** 1362.
Miller, J. E. (1965). *Biochem. Biophys. Res. Commun.* **19,** 335.
Robin, Y., and Viala, B. (1966). *Comp. Biochem. Physiol.* **18,** 405.
Ryley, J. F. (1952). *Biochem. J.* **52,** 438.
Seaman, G. R. (1955). *In* "Biochemistry and Physiology of Protozoa" (S. H. Hutner and A. Lwoff, eds.), Vol. 2, p. 91. Academic Press, New York.
Van Niel, C. B. (1949). *In* "Photosynthesis in Plants" (J. Frank and W. E. Loomis, eds.), p. 437. The Iowa State College Press, Ames, Iowa.
Warnock, L. G., and van Eys, J. (1962). *J. Cell. Comp. Physiol.* **60,** 53.
Williams, B. L., Goodwin, T. W., and Ryley, J. F. (1966). *J. Protozool.* **13,** 227.
Yamanaka, T., Nagata, Y., and Okunuki, K. (1968). *J. Biochem.* (*Tokyo*) **63,** 753.

Author Index

Numbers in *italics* refer to the pages on which the complete references are listed.

C

D

I

J

M

N

O

P

Q

R

S

T

Subject Index

A

B

C

D

E

F

G

H

I

J

K

L

M

T

U